2nd International Conference on Materials Engineering and Nanotechnology (ICMEN 2017)

IOP Conference Series: Materials Science and Engineering
Volume 205

Kuala Lumpur, Malaysia
12 – 14 May 2017

ISBN: 978-1-5108-4222-9
ISSN: 1757-8981

Printed from e-media with permission by:

Curran Associates, Inc.
57 Morehouse Lane
Red Hook, NY 12571

Printed by Curran Associates, Inc. (2017)

For permission requests, please contact the Institute of Physics at the address below.

Institute of Physics
Dirac House, Temple Back
Bristol BS1 6BE UK

Phone: 44 1 17 929 7481
Fax: 44 1 17 920 0979

techtracking@iop.org

Additional copies of this publication are available from:

Curran Associates, Inc.
57 Morehouse Lane
Red Hook, NY 12571 USA
Phone: 845-758-0400
Fax: 845-758-2633
Email: curran@proceedings.com
Web: www.proceedings.com

Table of contents

Volume 205

The 2nd International Conference on Materials Engineering and Nanotechnology
12–14 May 2017, Kuala Lumpur, Malaysia

Accepted papers received: 25 May 2017
Published online: 9 June 2017

Preface

Alloy behavior and mechanical properties

Performance analysis and experiment of cement

Material performance analysis and evaluation

Mechanical processing and methods of functional materials

Advanced material structure design and synthesis

Design and development of electronic devices

PREFACE

It is our great pleasure to introduce you the proceedings of 2017 2nd International Conference on Materials Engineering and Nanotechnology (ICMEN2017) held in Kuala Lumpur, Malaysia during May 12-14, 2017. ICMEN2017 is dedicated to issues related to Materials Engineering and Nanotechnology.

The conference program covered keynote, oral and poster presentations from scholars working in the areas of materials science and engineering to establish platforms for collaborative research projects in this field, and to find potential opportunities for international cooperation. This conference covered recent trends and progresses made in the field of materials engineering and nanotechnology.

The major goal and feature of the conference is to bring academic scientists, engineers, industry researchers together to exchange and share their experiences and research results, and discuss the practical challenges encountered and the solutions adopted. Professors from Malaysia and Turkey were invited to deliver keynote speeches regarding latest information in their respective expertise areas.

Distinguished scientists and experts in the area delivered the latest knowledge in their respective expertise fields. International scientific activities, like conference, symposium and workshop are big scientific platforms for knowledge exchange among the scientists from all over the world. International scientific activities are golden opportunities for the students, researchers and engineers to interact with the experts and specialists to get their advice or consultation on technical matters. The theme of this conference was the key issues associated with science and technology in this rapidly evolving field of research and to promote contact between basic researches and technological needs for real advanced materials and nanotechnology.

These proceedings present a selection from papers submitted to the conference from universities, research institutes and industries. All of the papers were subjected to peer-review by conference committee members and international reviewers. The papers selected depended on their quality and their relevancy to the conference. The volume tends to present to the readers the recent advances in the field of Materials Engineering and Nanotechnology. And various related areas, such as Nanotechnology, Nanomaterials, Nanoelectronics, Nanosystems, Nanomechanics, Nanomagnetism, etc.

We want to express our gratitude to all the members of conference committee, and reviewers who spared their valuable time, for their advice which have certainly helped to improve the quality, accuracy, relevance and the sincere efforts to maintain the quality of each paper selected for conference program and volume for publication. Also, we would like to thank all the authors who have contributed to this conference and also to the organizing committee, reviewers, speakers, chairpersons, sponsors and all the conference participants for their support to ICMEN 2017.

Prof. Dr. Mohd Hamdi Abdul Shukor, University of Malaya, Malaysia
Prof. Osman Adiguzel, Firat University, Department of Physics, Turkey
19. 05, 2017

IOP Conf. Series: Materials Science and Engineering **205** (2017) 011002 doi:10.1088/1757-899X/205/1/011002

INTERNATIONAL ADVISORY COMMITTEES

Prof. Chii-Ruey Lin, National Taipei University of Technology, Taiwan

CONFERENCE COMMITTEE CHAIRS

Prof. Dr. Mohd Hamdi Bin Abd Shukor, University of Malaya, Malaysia
Prof. Ir. Dr. Farid Nasir bin Haji Ani, Universiti Technologi Malaysia, Malaysia
Prof. Osman Adiguzel, Firat University, Department of Physics, Turkey

PROGRAM COMMITTEE CHAIRS

Prof. Arunachala Kannan, Arizona State University, USA
Prof. Ram Mohan, Joint School of Nanoscience and Nanoengineering, USA

INTERNATIONAL TECHNICAL COMMITTEES

Prof. Lee Yok Heng, Universiti Kebangsaan Malaysia, Malaysia
Prof. Iis Sopyan, International Islamic University Malaysia, Malaysia
Prof. Md. Mujibur Rahman, Universiti Tenaga Nasional, Malaysia
Prof. Ramesh Singh, University of Malaya, Malaysia
Dr. Li Yang, Xi'an Jiaotong Liverpool University, P. R. China
Assoc. Prof. Dr. Siti Norasmah Surip, Universiti Teknologi MARA Shah Alam, Malaysia
Prof. Mohamed Agouzoul, EMI RAbat & INSA Rouen france, Morocco
Prof. Vasily LUTSYK, Buryat State University, Russia
Mahendran Samykano, Universiti Malaysia, Malaysia
Prof. Hyun Do Yun, Chungnam National University, Dept. of Architectural Engineering, Daejeon, South Korea
Fatimah Noor, ITB, Indonesia
Assoc. Prof. Jun Xu, South China University of Technology, China
Assoc. Prof. Megat Ahmad Kamal Megat Hanafiah, Universitit Teknologi MARA Cawangan Pahang, Malaysia
Zhang Tingting, Dalian University of Technology, China
Dr. Zhenlin Wu, Dalian University of Technology, China
Asst. Prof. Kaykhosrow Khojier, Department of physics, Chalous branch, Islamic Azad University, Iran.
Prof. Ch. Sanjay, GITAM University Hyderabad, India
Dr. Ervina Efzan binti Mhd Noor, Multimedia University, Malaysia
Khadijah Khalid, Universitit Teknologi MARA Cawangan Pahang, Malaysia
Asst. Prof. Jeyashelly Andas, Universiti Teknologi MARA, Malaysia
Asst. Prof. Jully Tan, UCSI University, Malaysia
Chalongrat Daengngam, Prince of Songkla University, Thailand
Asst. Prof. Gobinda Gopal Khan, Tripura University (A Central University), Tripura, India
Asst. Prof. Perumal Elumalai, Pondicherry University, India

IOP Conf. Series: Materials Science and Engineering **205** (2017) 011002 doi:10.1088/1757-899X/205/1/011002

Asst. Prof. Karthik Silaipillayarputhur, King Faisal University, Saudi Arabia
Salmia Beddu, Universiti Tenaga Nasional, Malaysia
SITI KHOLIJAH BINTI ABDUL MUDALIP, Universiti Malaysia Pahang, Malaysia
Hilton Ahmad, UNIVERSITI TUN HUSSEIN ONN MALAYSIA, Malaysia

IOP Conf. Series: Materials Science and Engineering **205** (2017) 011003 doi:10.1088/1757-899X/205/1/011003

Peer review statement

All papers published in this volume of *IOP Conference Series: Materials Science and Engineering* have been peer reviewed through processes administered by the proceedings Editors. Reviews were conducted by expert referees to the professional and scientific standards expected of a proceedings journal published by IOP Publishing.

IOP Conf. Series: Materials Science and Engineering **205** (2017) 012001 doi:10.1088/1757-899X/205/1/012001

Hot Deformation Behavior of High Strength Low Alloy Steel by Thermo Mechanical Simulator and Finite Element Method

Kingkam W [1], Li N[1], Zhang H X[1, 2] and Zhao C Z[1, 2]

[1]College of Materials Science and Chemical Engineering, Harbin Engineering University, Harbin 150001, China
[2]Key Laboratory of Superlight Materials and Surface Technology, Ministry of Education
E-mail: zhaochengzhi@hrbeu.edu.cn

Abstract. The hot deformation behavior of HSLA steel was investigated by using a MMS-200 thermal mechanical machine at different conditions and with deformation temperature of 800-1100 °C and strain rate of 0.1-10 S^{-1}. FEM was analyzed the deformation characteristics of hot compression through Deform-3D software. It was discovered that the flow stress increases with increasing strain rate and decreasing temperature. The activation energy and stress exponent during hot deformation were calculated using hyperbolic sine constitutive equations. The result from the experiment represents the activation energy and stress exponent during hot compression of 222.256 kJ/mol. and 10.84. The prediction of distribution stress values from the constitutive equation in Deform-3D can be matched with the experimental results.

1. Introduction

High strength low alloy steel (HSLA) is a type of steel widely used for structural construction. Steels with low carbon content and very small additions of strong carbide or carbonitride forming elements such as Nb, V, and Ti to strengthening precipitation and grain refinement are called microalloyed or high strength low alloy steel [1]. These steels are much stronger and tougher than ordinary carbon steels, more ductile, highly formable, weldable and highly resistant to corrosion. Nowadays, HSLA steel has been improved with suitable chemical composition and thermo mechanical treatment in order to reinforce its application in heavy industries and automotive.

Recently, HSLA steel has been investigated for several features, including its hot deformation behaviors. Hot deformation process is necessary to improve the effectiveness of HSLA steel. With the increasing use of FEM (finite element method) to simulate the specimen behavior under the various parameter of compression test. The relationships between the constitutive equation and the relating process variables such as strain rate and temperature to the flow stress of the deforming material is required, and it is important to calculate the flow stress[2]. Flow stress can be defined as the resistance of a material against plastic deformation and it is expressed as a function of temperature, strain, and strain rate. Hence, FEM technique provides an effective approach for evaluating and determining the distribution and variation of thermo mechanical parameters in deformed specimens during hot working [3,4].

In this paper, the influence of characteristic hot deformation behavior of HSLA steel during compression test at different temperature and strain rate will be investigated. The general constitutive equations were used to determine the hot deformation constants of the material from experimental studies. In addition, the numerical analysis was carried out by Deform-3D simulation of the deformation characteristics in the hot compression process.

2. Experimental

The HSLA steel casting ingot was used for investigating this experiment. The chemical composition of HSLA steel is shown in Table 1.

Table 1. Chemical composition of High strength low alloy steels (wt%)

C	Si	Mn	Cr	Mo	Ni	Al	Cu	Nb	Ti	V
0.0031	0.151	0.766	0.190	0.136	0.517	0.0243	0.365	0.0408	0.0020	0.0145

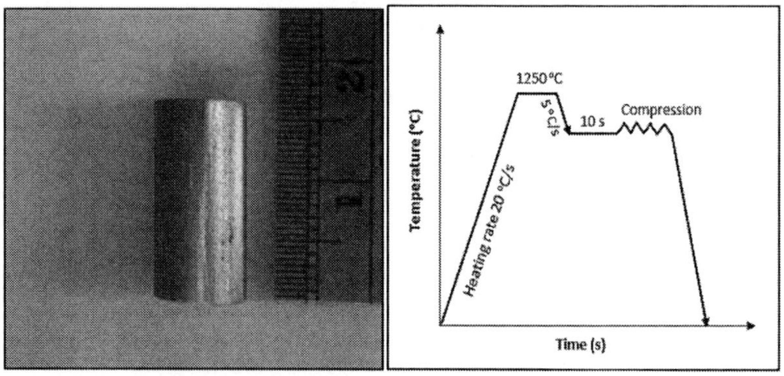

Figure 1. The cylindrical specimen and schematic of the hot compression test

Figure 2. MMS-200 thermal-mechanical simulator

The cylindrical specimens were machined with 8 mm in diameter and 15 mm in height. Hot compression test was performed by MMS-200 thermal-mechanical simulator. The specimens were heated from room temperature to 1250 °C at a heating rate of 20 °C/s for 5 min and then cooled to a deformation temperature with cooling rate 5 °C/s. After holding for 10 s. Compression deformation temperatures were conducted at 800, 900 and 1100 ℃ , respectively with different strain rates from 0.1, 1 to10 s^{-1}.The cylindrical specimens and schematic of hot compression experiment as shown in Figure.1 and Figure.3 shows the specimens after the hot compression test.

Finite element method (FEM) simulate the hot compression process was performed using the above true stress-strain data, and modify the true stress-strain data based on the difference value of resulting FEM load-displacement data and that of experimental. FEM to simulate the hot compression process was performed using the modified true stress-strain data again and recalculated the difference value of resulting the experimental data. Repeat the process till the entire match. Finite element simulation of the above hot compression has been performed using a commercial software package Deform-3D.

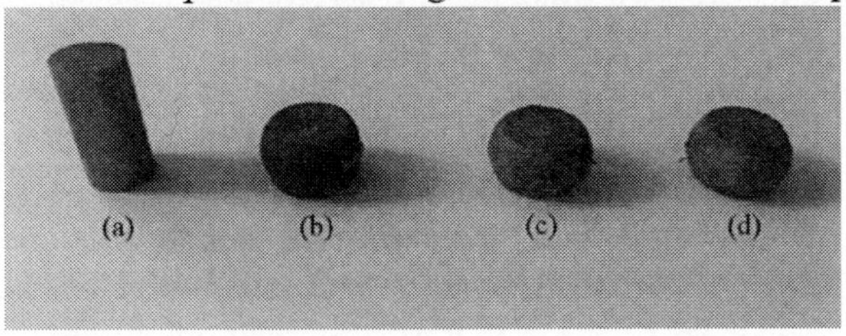

Figure 3. Specimens after hot compression test under different temperature at strain rate 1 s^{-1} (a) Original specimen, (b) 800 °C, (b) 900 °C and (c) 1100 °C

3. Results and Discussion

3.1 Stress-strain curves

The experimental studies of the true stress-strain curves for HSLA steel deformed at different deformation temperature and strain rate up to a height reduction of 50% are shown in figure.4.The flow stress curve shows the different strain rate at temperature 800 °C, it can be observed that the strain peaks increases with the elevating strain rate. Dynamic recrystallization phenomenon occurred at strain rate 0.1 s^{-1} and dynamic recovery occurred when the strain rate is 1 and 10 °C

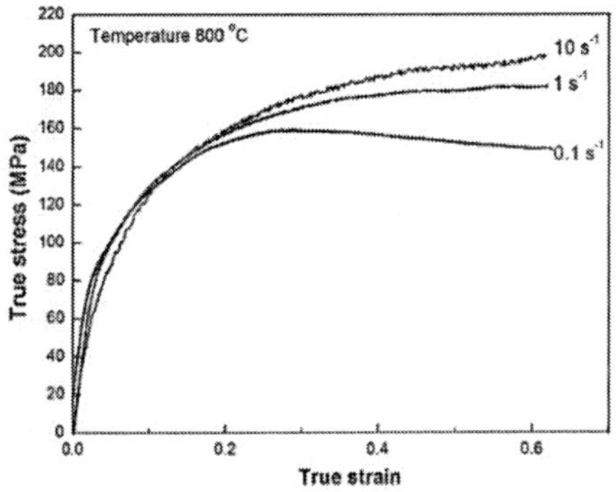

Figure 4. True stress-strain curve for different strain rate at deformation temperature 800 °C

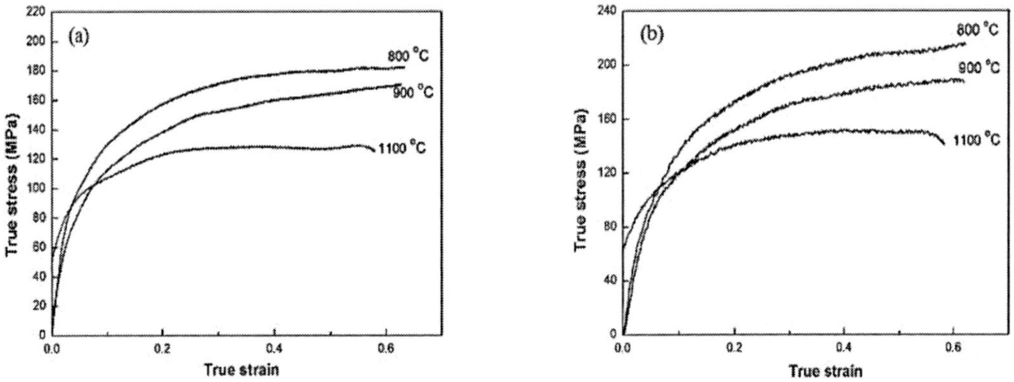

Figure 5. True stress-strain curve at different deformation temperature and strain rate (a) 1 s^{-1} and (b) 10 s^{-1}

Figure 5. shows the stress-strain curve at different deformation temperature and at strain rate 1 s^{-1} and 10 s^{-1}, it can be observed that the deformation temperature and strain rate have significant effect on the flow stress under all of the conditions tested. The true stress-strain curves exhibit a slightly lower value in flow-stress after obtaining the peak stress. It can also be observed that the flow stress and strain increases with decreasing deformation temperature because the dynamic softening dominates work hardening at higher temperatures [5, 6]. Subsequently, the flow stress curves continuously decreases until a balance with work hardening is achieved and shows dynamic softening. The stress value decreases with increasing deformation temperature, this is because the higher temperatures and lower strain rates provides a longer time for energy accumulation, higher mobilities at boundaries for the nucleation and dynamically recrystallized grains to grow [7]

3.2 Constitutive equation of hot deformation

Constitutive equations usually describe the relation between flow stress and flow strain under different strain rate and temperature. The flow stress-strain data obtained from experimental and can be used to determine the materials constantly. The relationship between the flow stress (σ), temperature (T) and strain rate $(\dot{\varepsilon})$ is expressed by hyperbolic sine function form of the Arrhenius type equation. The effects of strain rate and temperature plastic deformation behavior could be demonstrated by Zener-Holloman parameter [8-10]. In hot deformation, several constitutive equations have commonly been applied by Eq.(1) and (2).

$$Z = \dot{\varepsilon} \exp\left(\frac{Q}{RT}\right) \tag{1}$$

$$\dot{\varepsilon} = AF(\sigma)\exp\left(-\frac{Q}{RT}\right) \tag{2}$$

$$\text{Where, } F(\sigma) = \begin{cases} \sigma^n & \alpha\sigma < 8 \\ \exp(\beta\sigma) & \alpha\sigma > 1.2 < 0.8 \\ [\sinh(\alpha\sigma)]^n & all\,\sigma \end{cases}$$

where $\dot{\varepsilon}$ is the strain rate (s^{-1}); Z is the Zener-Holloman parameter; Q is the activation energy of hot deformation ($kJ \cdot mol^{-1}$); R is the gas constant (8.314 $J \cdot mol^{-1} \cdot K^{-1}$), T is the absolute temperature (K), σ is the peak stress (MPa) A, α, and n are the materials constants.

The correlation between strain rate and temperature can be expressed by parameter Z as follows.

$$Z = \dot{\varepsilon} \exp\left(\frac{Q}{RT}\right) = A[\sinh(\alpha\sigma)]^n \tag{3}$$

Where, Z is the factor of strain rate of compensated with temperature, $\alpha = \beta / n$ α MPa^{-1}, n and β are the constant.

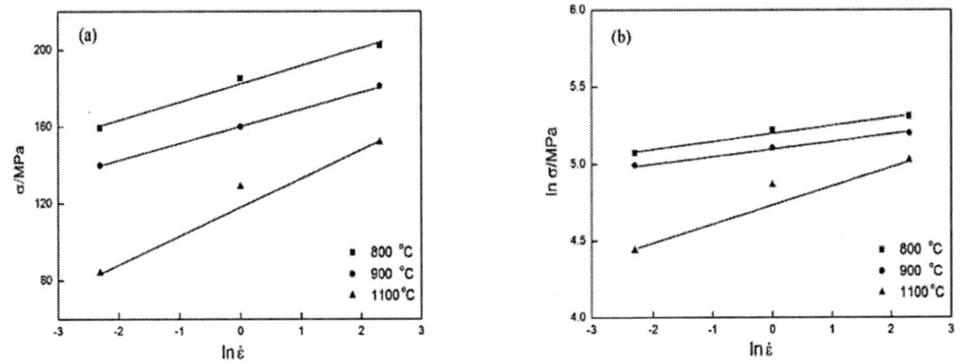

Figure 6. Relationship between strain rate and peak stress (a) ln $\dot{\varepsilon}$-σ and (b) ln $\dot{\varepsilon}$- ln σ

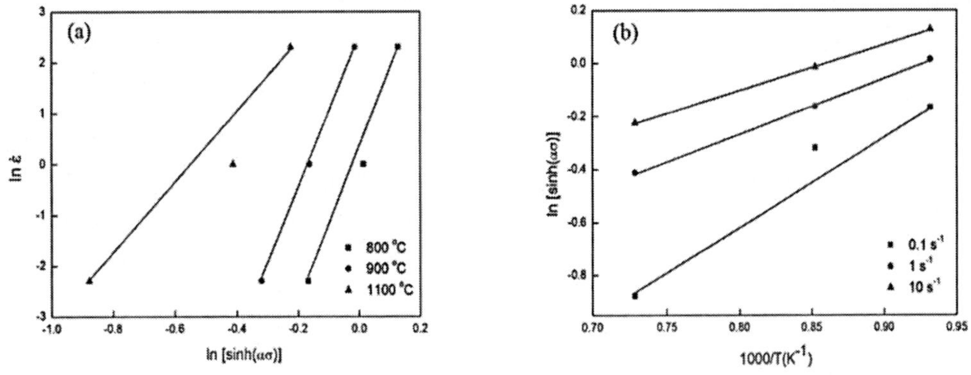

Figure 7. Relationship between (a) $\ln[\sinh(\alpha\sigma)]$- ln $\dot{\varepsilon}$ and (b) $\ln[\sinh(\alpha\sigma)]$-T

The ln $\dot{\varepsilon}$-ln σ and ln $\dot{\varepsilon}$-σ could be obtained when taken logarithm on Eq. (3), especially the temperature is constant. For determining the values of Q, A and α, two parameters n_1 and β was given by Eq.(4) and Eq. (5).

$$n_1 = \left(\frac{\partial \ln \dot{\varepsilon}}{\partial \ln \sigma}\right)_T \tag{4}$$

$$\beta = \left(\frac{\partial \ln \dot{\varepsilon}}{\partial \sigma}\right)_T \tag{5}$$

$$n = \left(\frac{\partial \ln \dot{\varepsilon}}{\partial \ln[\sinh(\alpha\sigma)]}\right)_T \tag{6}$$

According to Eq. (4) and (5), the linear relationship between ln $\dot{\varepsilon}$-ln σ and ln $\dot{\varepsilon}$-σ are shown in Figure.6. The slope are calculated by using the least squares method of linear regression and take slopes reciprocal and averaging with result of value $\beta = 0.0771$ MPa^{-1}, $n_1 = 16.0083$ and $\alpha = 0.0048$. According to Eq. (6), the slope of plots of $\ln[\sinh(\alpha\sigma)]$- ln $\dot{\varepsilon}$ as shown in figure.7 (a) can be obtained the value of n. The average value of stress exponent (n) was defined as 10.8482.

For the all stress values, Eq. (2) can be given by Eq. (7) and then taking the logarithm of both sides and giving partial different the expression for the activation energy at a constant strain rate, for the result can be obtained in Eq.(8)

$$\dot{\varepsilon} = A\left(\sinh(\alpha\sigma)^n \exp\left(-\frac{Q}{RT}\right)\right) \tag{7}$$

$$Q = Rn\left(\frac{[\partial \sinh(\alpha\sigma)]}{\partial(1/T)}\right)_{\dot{\varepsilon}} \tag{8}$$

The value of Q can be obtained from the slope of every plot of $\ln[\sinh(\alpha\sigma)]$-T as shown in figure.7 (b). For the average value of activation energy (Q) = 222.256 kJ/mol and A = 3.889x10^{12}

Table 2 Parameter of constitutive equation for experimental of HSLA steel

Material constants	Q (kJ/mol)	N	α	β (MPa^{-1})	A (S^{-1})
value	222.256	10.8482	0.0048	0.0771	3.889x10^{12}

3.3 FEM Simulation

The effects of deformation temperature on the stress-strain distribution of HSLS steel are described using the finite element method (FEM). MMS-200 thermal mechanical simulator was used to acquire accurate flow stress curves of HSLA steel. The dimension of the specimens is equally considered in the experiment. The flow stress curves obtained from the experimental studies were functions of temperature, stress and strain rate. These curves were imported into DEFORM-3D for the simulations procedure.

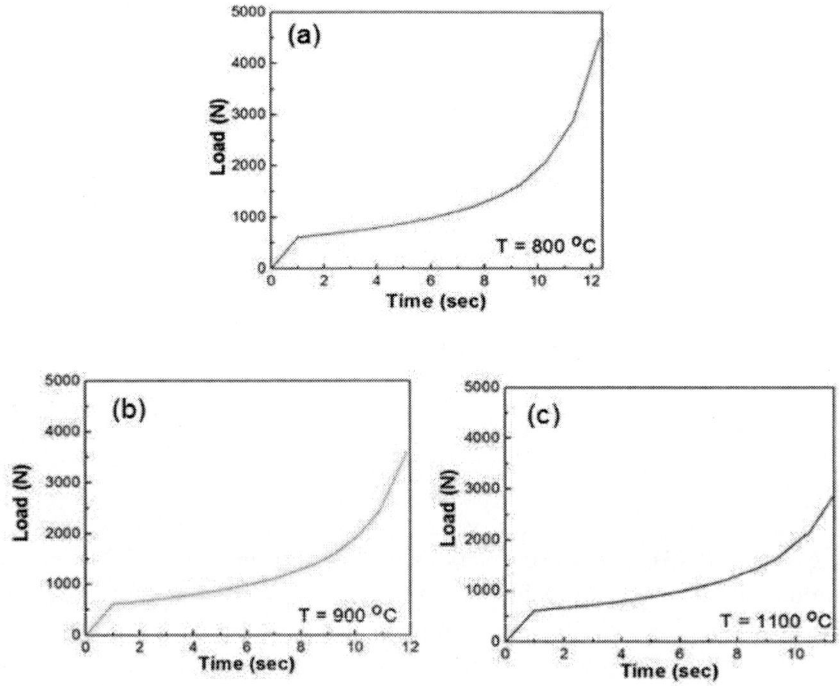

Figure 8. Load-time curve under different temperature at strain rate 1 s^{-1} (a) 800 °C, (b) 900 °C, (c) 1100 °C

The effect of deformation temperature on the load-time curves at strain rate 1 s^{-1} is shown in figure 8. It can be observed that the variation in load–time curves, which depends on the stress, decreases with the increasing of deformation temperatures. Similarly, the difference in the load–time curves is lower at higher temperature and during the distribution of the effective stress under different temperature; the maximum effective stress changes with the variations in deformation temperatures. When the deformation temperature is lower, the deformation of specimens strain value increases. For the high-temperature conditions, the effective stress distribution value is lower and it shows that the deformation of areas is relatively large as shown in figure 9. The value of maximum effective stress at the temperature 800°C, 900°C and 1100 °C are 194.98 MPa, 159.77 MPa and 113.62 MPa respectively.

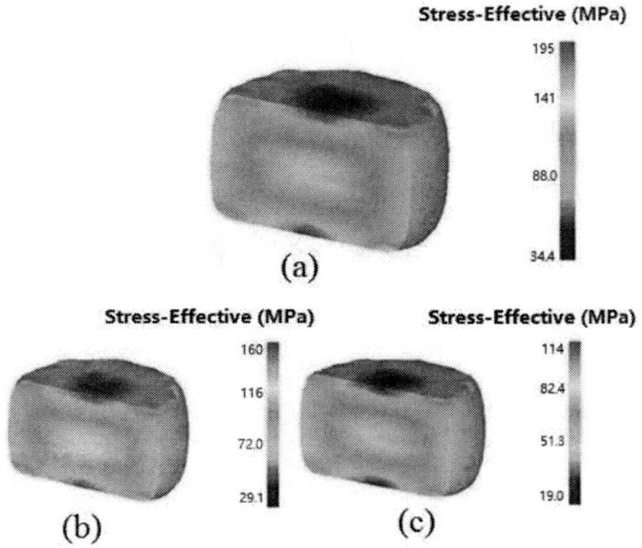

Figure 9. Effect of temperature on the stress distribution at strain rate 1 s^{-1} and height reduction of 50% (a) 800 °C, (b) 900 °C, (c) 1100 °C

4. Conclusion

1. The true stress-strain curves of HSLA steel shows different strain rate at constant deformation temperature. Dynamic recrystallization characteristic surfaces at low strain rate when the higher strain rate dynamic recovery has occurred and stress and strain peaks increases with increasing strain rate and decreasing deformation temperature.
2. The constitutive equation of hot deformation for HSLA steel under the various conditions was calculated the activation energy (Q) by using hyperbolic sine function law. The result from regression analysis shows that the activation energy of hot compression from experimental studies was 222.256 kJ/mol.
3. The simulation of the influence of the strain rate on the state of strain and stress shows that the variation in load–time curves, which depend on the stress, decreases with the increasing deformation temperatures and the distribution of the effective stress under different temperature is similar to the experimental. The prediction of flow stress values from the constitutive equation in Deform-3D can be matched with the results of the experimental.

Acknowledgment

This paper is funded by the International Exchange Program of Harbin Engineering University for Innovation-oriented Talents Cultivation and Harbin Science and Technology Achievement Transformation Project (2013DB4AG009).

Reference

[1] Brick R M, Pense A W and R B Gordon 1977 *Structure and Properties of Engineering Materials*, 4th Ed. McGraw-Hill NY USA.
[2] Chen X and Huang Y 2015 *J. Alloys Compd* vol 619 p 564–571.
[3] Park N K, Yeom J T and Na YS 2002 *J. Mater. Process. Technol* vol 130–131 p 540–545.
[4] Goetz R L and Semiatin S L 2001 *J. Mater. Eng. Perform* vol 10 no 6 p 710–717.
[5] Deng Y, Yin Z, and Huang 2011 *J. Mater. Sci. Eng. A* vol 528 no 3 p 1780–1786.
[6] Li J, Li F, Xue F, Cai J, and Chen B 2012 *Mater. Des* vol 37 pp 491–499.
[7] Xu D, Zhu M, Tang Z and Sun C 2013 *J. Wuhan Univ.Technol.-Mater Sci Ed* vol 28 no 4 p 819–824.
[8] Zener C and Hollomon J H 1944 *J. Appl. Phys* vol 15 no 1 p 22–32.
[9] Saravanan L and Senthilvelan T 2015 *J Mater restechnol* vol 5 no 1 p 21–28.
[10] Lin Y C, Chen M S and Zhong J 2008 *J. Mater. Process. Technol* vol 205 no 1–3 p 308–315.

IOP Conf. Series: Materials Science and Engineering **205** (2017) 012002 doi:10.1088/1757-899X/205/1/012002

Influence of Nano-3%Al_2O_3 on the Properties of Low Temperature Sn-58Bi (SB) Lead-free Solder Alloy

S Amares[1, 4], **M N Ervina Efzan**[2], **Rajkumar Durairaj**[3] and **Aliasghar Niakan**[1]

[1]Faculty of Engineering and Built Environment, SEGi University, No. 9, Jalan Teknologi, Taman Sains Selangor, Kota Damansara PJU 5, 47810 Petaling Jaya, Selangor, Malaysia.
[2]Faculty of Engineering and Technology, Multimedia University, 75450 Ayer Keroh, Melaka Malaysia.
[3]Lee Kong Chian Faculty of Engineering and Science, Universiti Tunku Abdul Rahman, Jalan Sungai Long, Bandar Sungai Long, 43000 Kajang, Selangor, Malaysia.

amaressinghgill@segi.edu.my

Abstract. This work studies the melting temperature, wettability, metallurgical and hardness properties of the Sn-58Bi (SB) lead-free solder alloy incorporated with nano-3%Al_2O_3. The melting temperature was observed at 143.44°C upon the additions of the nano-3%Al_2O_3 with a low contact angle of 20.4°. A well-distributed microstructure with narrower lamellar structure and finer intermetallic compounds and Sn grains was detected for the nano-3%Al_2O_3 added SB solder alloy. Hardness evaluation based on the Vickers hardness value was as high as 17.1Hv. Overall, the Sn-58Bi + 3% Al_2O_3 solder alloy appears to harvest beneficial results for these properties and can be used as potential replacement in the current electronic packaging industry.

1. Introduction

Ultimately, efficiency of a solder alloy depends on the alloying process and examples such as Sn-Pb, Sn-Bi and Sn-Ag-Cu are a mixture of one or more elements producing a solder alloy. Alloying seems to boost the performance of a solder alloy by enhancing its property such as lowering the melting temperature (e.g. Sn-Pb, Sn-Bi), producing high hardness (e.g. Sn-Zn-Bi) and better wetting for better joint property [1]. Among many solder alloys, the Sn-Bi (SB) solder alloys are pointed out to be the potential candidate to be used in the electronic packaging industry. Studies by [2] and [3] agrees to this point. However, the SB solder alloys has drawback such as low mechanical properties and these properties are key to ensure the functional integrity of the electrical component [4]. Therefore, introductions of nanoparticle appears to overcome the concern [5]. Parallel to that, the melting, mechanical and metallurgical properties of the low melting temperature Sn-58Bi (SB) added with 3% Al_2O_3 nanoparticles solder alloy were studied in this research. The results were analysed and further discussed to provide information in this area of studies.

2. Experimental Setup

The Sn-58Bi (SB) + 3% Al_2O_3 solder alloy was prepared from 8.4g of tin (99.9%, Alfa Aesar), 11.6g of bismuth (99.9%, Alfa Aesar) and 3% of Al_2O_3 nanoparticles which then were mechanically mixed by melting in a furnace at 600°C for 1 hour to ensure a homogenous mixture. To enable a proper mixing of the nanoparticles, the solder alloy was re-melted using a hot plate at 300°C. Figure 1 confirms that the mechanical mixing method effectively incorporated the Al_2O_3 nanoparticles due to the presence of the aluminium (Al) and oxygen (O) as labelled. The Sn-58Bi (SB) + 3%Al_2O_3 solder alloy were then compressed into billets for the melting and hardness test. Prior to the test, the billets were polished and etched in Nital for 60 seconds and observed under the Optical Microscope for microstructural study. The wettability test was conducted by soldering the Sn-58Bi + 3% Al_2O_3 solder alloy on to the Copper (Cu) substrate and the contact angle was measured using the VIS Pro software. Melting temperature was studied by using the Differential Scanning Calorimetry (DSC) test. Hardness test meanwhile was conducted based on Vickers hardness with load of 1kN applied to the Sn-58Bi + 3% Al_2O_3 solder alloy.

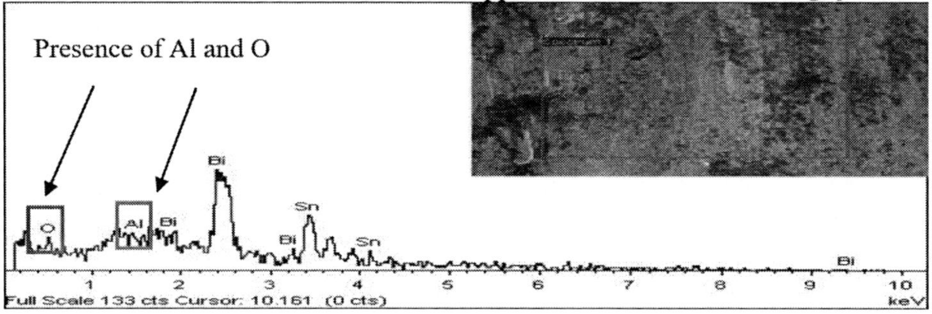

Figure 1. Elemental analysis of 3% Al_2O_3 nanoparticles in SB solder alloy.

3. Melting Temperature

The DSC graphs for the Sn-58Bi + 3% Al_2O_3 solder alloy shows a sharp endothermic peak (figure 2) with the start of melting (solidus temperature, T_S) at 139.43°C, peak temperature which is the melting temperature, T_M at 143.44°C, and the liquidus temperature, T_L at 147.29°C indication of the point solder being fully melted. Comparing the bare Sn-Bi solder alloy (T_M=138°C) [6], a slight increment of 3% in the melting temperature was observed. Apparently, the Al_2O_3 nanoparticles retains its solubility limit in the molten solder alloy with no diffusion of the Al_2O_3 nanoparticles that could affect the melting temperature or in other words, the Al_2O_3 nanoparticles acts as discrete particles [24]. Hence, the slight increase in the melting temperature occurs due to minor local dissolution of the Al_2O_3 that has high melting temperature of 2072°C [7]. Similar observation noted by [8] upon adding zirconia, ZrO_2 as the ZrO_2 elevates the melting temperature of Sn-3.0Ag-0.5Cu solder system. Another reason is that the Al_2O_3 nanoparticles tends to gather at the surface of the molten solder alloy due to its high energy [9] and this increases the energy of the solder to melt and consequently making the melting temperature to increase [10]. Yet, such increase as detected is not a major concern as the increment is minimal and the melting range still falls under the low temperature solder alloys (< 150°C) [11]. In fact, the Sn-58Bi + 3% Al_2O_3 solder alloy has lower temperature compared to many other solders, Sn-Pb (T_M=183°C) [12], Sn-0.7Cu (T_M=227°C) [13], Sn-Ag-Cu (T_M=217°C) [14] and other solder alloys.

4. Wettability

The wettability is a crucial aspect that displays the integrity of the solder to the substrate and is studied based on the contact angle between the solder and substrate [15]. Lower contact angles are very much desired because it is known to avoid crack propagations [16]. The SB + 3 % of Al_2O_3 solder alloy was soldered at temperature of 250°C, the soldering temperature in the electronic industry [17]and the contact angle was measured to be 20.4°, table 1. Compared to the traditional Sn-Pb solder alloy producing contact angle of 8° [18], the SB + 3 % of Al_2O_3 solder alloy has 40% higher contact angle.

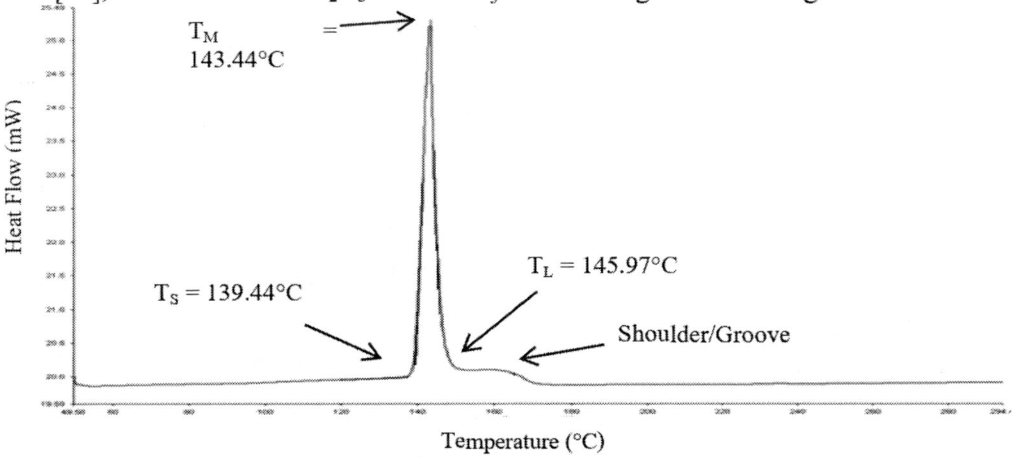

Figure 2. DSC curve of SB + 3% Al_2O_3

The presence Al_2O_3 nanoparticles will increase the melting temperature as pointed earlier [19] and disables a better spreading as a result in reduction in the fluidity for the solder alloy [7]. On top of that, the Al_2O_3 nanoparticles will increase the activation energy, leading to harder spreading of the molten SB solder alloy. The study of [20] noted similar reason. Additionally, typical characteristics of discrete nanoparticles are known to gather at the leading edge of the molten solder that then decreases the fluidity and increases the viscosity due to increase in surface tension [21]. Despite that, the SB + 3% Al_2O_3 solder alloy produces lower contact angle than most solder alloys, table 1. The contact angle of the SB + 3% Al_2O_3 solder alloy also falls within the desired range of contact angle (<30°) as verified by [22].

Table 1. Contact angle of solder alloys.

Solder Alloy	Average Contact Angle (°)	References
SB + 3%Al_2O_3	20.4	This Study
Sn-40Pb	8	[23]
Sn–3.5Ag	38	[24]
Sn–0.7Cu	35	
Bi-40Sn-2In	31	[25]
Sn-3.5Ag-0.7Cu	48	[26]

5. Microstructure Characteristics.

The additions of 3% Al_2O_3 nanoparticles in the SB solder alloy produces the microstructures as shown in figure 3. Result prevails there are light and dark phase identified as the bismuth precipitations and Sn-matrices respectively. The identification of the phases was similarly noted in the study of [27]. Upon additions of the nanoparticles, the grain sizes of the Sn-matrices were finely dispersed with the nearer gap between the alternating lamellar structures (black dotted region), figure 3. The existence of the Al_2O_3 nanoparticles will accumulate at the grain boundary of the solder and enables them to act as a site for nucleation for the primary Sn matrices and produce smaller and finer Sn matrices [28]. The theory of adsorption stated by [29] explains that the plane with the maximum surface tension grows fastest with an increasing adsorption element, the Al_2O_3 nanoparticles in this case for this study. However, with accommodation of many Al_2O_3 nanoparticles, the surface energy will be reduced together with the growth velocity [8], ensuring in finer microstructure as shown figure 3. The production of finer grain is also due to the heterogeneous nucleation site and impeded phase boundary due to the presence of the Al_2O_3 nanoparticles, an explanation likewise in the study of [15].

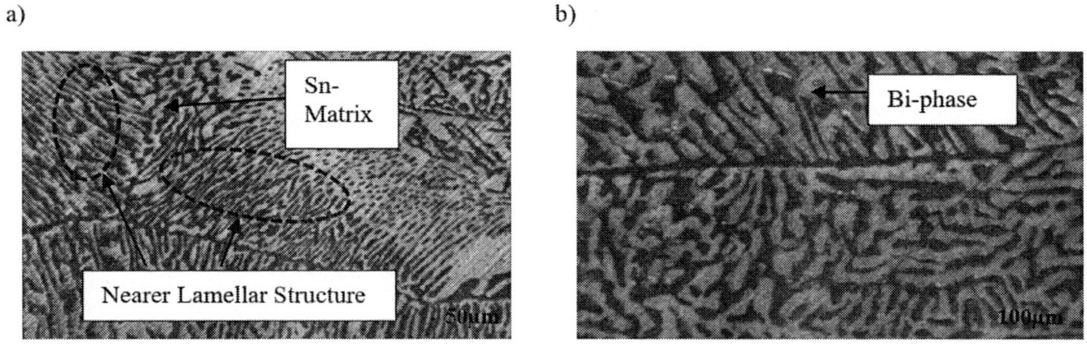

Figure 3. Microstructures of SB + 3% Al_2O_3 solder alloy, a) 50× and b) 100× magnification.

6. Hardness

The Vickers hardness for the SB + 3% of Al_2O_3 solder alloy was taken an average of 17.1Hv was noted based on five different indentations. Concerning the hardness, it is closely related to the well-defined microstructure and the presence of finer grains together with nanoparticles contributing to resisting deformations [30]. In this study, the reason behind the increase in hardness of the SB + 3% of Al_2O_3 solder alloy are due to the microstructure and the existence of the Al_2O_3 nanoparticles as well. The Al_2O_3 nanoparticles additions modifies the microstructure of the SB solder alloy and produces finer and nearer alternated lamellar structure together with a small grain production presented in figure 3. Consequently, these characteristics help the solder alloy to resist deformation by providing more stress needed to disrupt the microstructure configurations. Comparable explanation was confirmed by [4]. Apart from that, the Al_2O_3 nanoparticles contributes to increase in hardness by standing-in as hard particles accommodating in between the grain boundaries [31]. As when a loads are applied, these nanoparticles at the grain boundaries will be barrier for the dislocation to occur and directly increase the load needed to overcome the nanoparticles [31].

7. Conclusion

The objective of this study is to produce and analysis the characteristics of the Sn-58Bi + 3% Al_2O_3 solder alloy based key properties required in a solder allot. The Sn-58Bi + 3% Al_2O_3 solder alloy seems to have a low melting temperature of 143.44°C which is relatively lower than the Sn-Pb and the Sn-Ag-Cu solder alloy. At the same time, a lower contact angle portrays a good wettability characteristics of the solder. The solder alloy also produces a high hardness of averaged 17.1Hv which suggests that it offers resistance for deformation due to the well-defined microstructure together with the presence of 3% Al_2O_3 nanoparticles. Not many studies are conducted in incorporating nanoparticles to boost the properties, which is the objective of this research. Moreover, based on the study and results achieved, the Sn-58Bi + 3% Al_2O_3

nanoparticles solder alloy could be an alternative for the problem faced in the electronic packaging industry.

8. References

[1] Gao F, Qu J and Takemoto T 2010 *Journal of Electronic Materials Journal* **39** 426

[2] Xu DX, Lei PY, Zhi DX, Fu G and Yao WS 2008 *Journal of Electronic Materials Journal* **37** 125

[3] Li J, Mannan S, Clode M, Whalley D and Hutt D 2006 *Acta Materialia* **54** 2907

[4] Shen J and Chan Y 2009 *Microelectronics Reliability* **49** 223

[5] Tsao L, Chang S, Lee C, Sun W and Huang C 2010 *Materials and Design* **31** 4831

[6] Abtew M and Selvaduray G 2000 *Materials Science and Engineering* **27** 95

[7] Lee Y, Kim I, Kim E, Lee J and Moon J 2010 *Electronics Packaging Technology Conference (EPTC) vol 10* (Singapore IEEE) p 1

[8] Gain AK, Chan Y and Yung W 2011 *Microelectronics Reliability* **51** 2306

[9] Gain AK and Chan Y 2012 *Intermetallics* **29** 48

[10] Salleh MM, Al Bakri MA, Zan@Hazizi M, Somidin F, Alui NF and Ahmad ZA 2012 *Materials Science and Engineering* **556** 633

[11] Kanlayasiri K, Mongkolwongrojn M and Ariga T 2009 *Journal of Alloys and Compounds* **485** 225

[12] Yeo H and Han K 2009 *Journal of Alloys and Compounds* **477** 278

[13] Satyanarayan and Prabhu K 2011 *Advances in Colloid and Interface Science* **166** 87

[14] Chuang C, Tsao L, Lin H and Feng L 2012 *Materials Science and Engineering* **558** 478

[15] Ervina Efzan M, Amares S and Yap T 2013 *Soldering and Surface Mount Technology* **25** 229

[16] Ervina Efzan M and Tan S 2013 *Applied Mechanics and Materials*, **315** 675

[17] Amares S and Ervina Efzan MN 2015 *MATEC Web of Conferences* **27** 10

[18] Shalaby RM 2013 *International Journal of Physics and Research* **3** 1

[19] Chang S, Jain C, Chuang T, Feng L and Tsao L 2011 *Materials & Design* **32** 4720

[20] An B and Wu CL 2007 *International Conferecne Electronics Packaging Technology Conference (ICEPT)* vol 10 (China IEEE) p 1

[21] Nai S, Wei J and Gupta M 2006 *Thin Solid Films* **504** 401

[22] Chung C, Mustapha F, Hua F and Aspandiar R. 2002 An assessment of lead free solder (Sn3.7Ag0.8Cu) wettability *Electronics Packaging Technology Conference* vol 10 (Singapore IEEE)

[23] Xiaorui Z, Hiroyuki M, Fumitaka T and Zhangfu Y 2012 *Materials Transactions* **53** 926

[24] Likun Z, Zhangfu Y, Hongxin Z and Xiaorui Z 2009 *Materials Letters* **63** 2067

[25] Shalaby RM 2013 *Materials Science and Engineering* **560** 86

[26] Law C, Wu C, Yu D, Wang L and Lai J 2006 *Journal of Electronic Materials* **35** 89

[27] Miao HW and Duh JG 2001 *Materials Chemistry and Physics* **71** 255

[28] Tsao L, Wu R, Cheng TH, Fan KH and Chen R 2013. *Materials & Design* **50** 774

[29] Hu X, Xu S, Yang Y, Chen Z and Chan Y 2014 *Materials Science and Engineering* **600** 67

[30] Haseeb A, Leong Y and Arafat M 2014 *Intermetallics* **54** 86

[31] El-Daly A, Al-Ganainy G, Fawzy A and Younis M 2014. *Materials and Design* **55** 837

Microstructure and Oxidation Behavior of Cr/Mo Modified TiAl Alloy Containing High Nb

Zhu-Hang Jiang[1],Cheng-Zhi Zhao[1*], Wen-di Li[1], He-Xin Zhang [1*]

Institute for metallic meterials, College of Materials Science and Chemical Engineering, Harbin Engineering University, Harbin 150001, P.R.China
This paper is funded by the International Exchange Program of Harbin Engineering University for Innovation-oriented Talents Cultivation

Email: zhaochengzhi@hrbeu.edu.cn

Abstract. In this paper, the microstructure and oxidation behaviour of Ti-45Al-8Nb-0.2Si-0.5W-0.8B, Ti-45Al-8Nb-2Cr-0.2Si-0.5W-0.8B, Ti-45Al-8Nb-2Mo-0.2Si-0.5W-0.8B, Ti-45Al-8Nb-2Cr-2Mo-0.2Si-0.5W-0.8B were studied.The alloys were designed and fabricated via vacuum arc melting in the protection of argon shield. The oxidation experiments were carried out at 1073K for 200h in laboratory air. Microstructure evolution, elemental analysis as well as the composition distribution of the oxide scale were performed using scanning electron microscopy (SEM), energy dispersive spectroscopy (EDS) and X-ray diffraction (XRD) technique, respectively. The results showed that the microstructure could be refined by adding of Cr and Mo to a smaller grain size.The heat treatment has a good impact on composition homogenization as well as the generation of γ-TiAl phase. The oxidation test shows that Cr and Mo modified alloy cannot improve the oxidation resistance of the alloy. The 0Cr2Mo alloy shows the poorest oxidation resistance in the three , while the Cr modified alloy can apparently slows down the oxidation rate after 50h, which owing to the formation of a compact oxidation scale.

1. Introduction

Lightweight superalloys are attracting attention worldwide in the manufacturing fields of aerospace, aircraft and high-speed transportation as high temperature structural material. Owing to the low density , excellent specific strength, high specific stiffness, good high temperature oxidation and good creep resistance[1,2],TiAl-based alloys are becoming a potential candidate of high temperature structual material that services as low-level turbine and exhaust valve.The high Nb containing TiAl based alloy exhibits superior creep properties and high temperature oxidation resistance compared to normal TiAl based alloy over 973K[1~8].

TiAl-based alloys have high mechanical strength both at room and high temperatures because of their ordered structure. Single phase TiAl-based alloy is poor in room temperature ductility. But with the presence of the second α_2 phase, it changes to near gamma phase which shows a better ductility and optimizes the microstructure. Alloying elements Cr, B, Si and W are added into the TiAl alloy for solid solution strengthening. It's found out that over 5 atom percentage of Nb can significantly improve the high temperature oxidation resistence[9]. The effect of trace alloying element like Cr, Mo and V on oxidation resistance of these elements have not been figured out clearly yet[10-13].

Our work meant to figure out the microstructure evolution and oxidation resistance of the alloys with the addition of Cr and Mo. To this end, four high Nb containing TiAl alloy was designed and fabricated with different atomic fractions of Cr and Mo via vacuum arc melting in the protection of argon, and then the samples were post heat-treated in order to find out the microstructure refinement effect. The ingots then past an isothermal oxidation test. The as-cast and heat-treated microstructure, oxidation scale morphology, oxidation section composition are characterized employing scanning electron microscopy (SEM), energy dispersive spectroscopy (EDS) and X-ray diffraction (XRD) technique.

2. Materials and methods

2.1. Specimen preparation

Four novel designed alloys Ti-45Al-8Nb-0.2Si-0.5W-0.8B, Ti-45Al-8Nb-2Cr-0.2Si-0.5W-0.8B, Ti-45Al-8Nb-2Mo-0.2Si-0.5W-0.8B and Ti-45Al-8Nb-2Cr-2Mo-0.2Si-0.5W-0.8B are namely 0Cr0Mo, 2Cr0Mo, 0Cr2Mo and 2Cr2Mo alloy respectively. Pure Al(≥99.9%), titanium sponge(≥99.7%), pure niobium(≥99.8%), Cr(≥99.98%), Mo(≥99.7%), W(≥99.95%), and AlB alloy(2.5%~3.5% mt of B) are adopted as raw materials and get cleaned up before melting. Vacuum melting was applied to cast the materials into button ingots in a non-self-consumed tungsten electrode argon arc vacuum furnace. Before melting, the furnace was firstly vacuumized to 0.005Pa, then argon was used as a protective gas and pressure in furnace finally stabilized at 0.03Pa. The melting process was repeated four times for material homogenization through the overturning platform. After the melting process, materials were cooled down in water cold copper crucible into button ingots of around 40g. The ingots were cut into small specimen with the dimension of 10mm×10mm×1mm after homogenization treatment in α-phase region(1473K) for 24h. Then specimen experienced a cyclic heat treatment, which is a three-time heat treatment in 1423K for 4h with air cooling. After the heat treatment, the ingots were polished up using 1500 SiC water proof abrasive papers before cleaned in acetone.

2.2. Isothermal oxidation test and microstructure observation

The specimens went through an isothermal oxidation text which was carried out at 1073K for 200h at atmospheric pressure using a chamber electric furnace. The specimens were moved out of the furnace after every certain fixed period of time. Everytime after cooled down, the increased weights would be measured by electric balance (error range ±0.01mg) after cooling down to room temperature. In order to minimize data error, two specimens at least were tested at a time.

The microstructure evolution were observed by back-scattered scanning electron microscope (SEM) before and after the mentioned heat treatment in 2.1. The oxidation samples were cut perpendicular to the surface of oxide and were polished by SiC papers up using 2000 and polished examined by SEM in purpose of observing oxidation section. And for oxidation surface observation, no special treatment are needed except keeping it clean.In order to find out the structure of oxidation scales, line scanning is used, as well as energy dispersive spectroscopy (EDS) analysis for oxide layer composition analysis. A PANalytical X'Pert diffractometer was applied to identify the phase, conditioning Cu Kα radiation under 40 kV and 40 mA, the scanning range (2θ) was from 10°-90°, and step size was 0.017°/step.

3. Results and discussion

Fig. 1 shows the microstructure of four alloys both in as-cast and heat-treated condition.A homogenization treatment(1473K for 24h) and cylic heat treatment(three-time 4h 1423K air cooling) is applied on every alloy.It is noted that there are two typical microstructures,which are fully lamellar microstructure in **Fig. 1(a)** and near fully lamellar microstructure in others. Fully lamellar microstructure is composed of alternating lamellae of α_2-Ti_3Al and γ-TiAl and is benificial to the fracture toughness and high-temperature strength properties of TiAl-based alloys, but the grain size and the inhomogeneous distribution in as-cast condition caused poor ductility in room temperature. Appropriate heat treatments can refine the phase distribution and grain size[14-16]. **Fig. 1(a)** and **Fig. 1(b)** shows that heat treatment leads to the appearance of single γ phase due to a transformation from high-temperature α to highly faulted γ phase which commonly occurs in rapid cooling process that suppresses diffusion or a result of adding of alloying elements. The microstructure that single γ phase appear in grain boundery of α_2-γ phase is what is called near fully lamellar microstructure. The near fully lamellar microstructure with more γ phase in the grain boundry can can lead to more Al_2O_3

protective oxidition scale during oxidation process because γ-TiAl phase has more Al volume fraction and better oxidation resistance than α_2-Ti_3Al phase. Adding of Cr and Mo helps the rising of single γ phase of the alloys, but as can be seen in **Fig. 1** that single γ phase takes up more volume fraction in Fig. 1(d) than Fig. 1(f), heat-treatment is more effective on Cr modified alloy than Mo modified alloy since more single γ phase means a better transformation from α_2 to γ phase. Furthermore, Cr and Mo both have very good grain refining effect that lead to smaller grain size and thinner interlamellar spacing. The three alloys went through the isothermal oxidation.

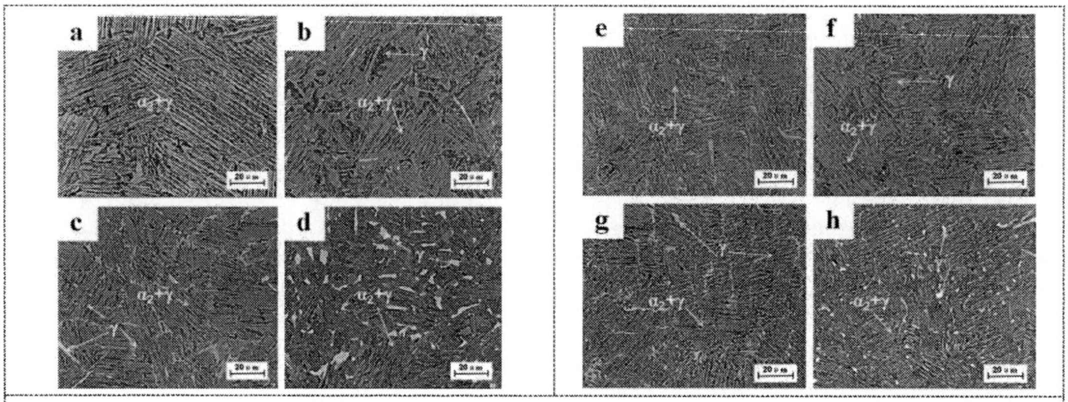

Figure 1. Back-scattered SEM metallograph of (a)(b)0Cr0Mo alloy, (c)(d)2Cr0Mo alloy, (e)(f)0Cr2Mo alloy, (g)(h)2Cr2Mo alloy(figure(a),(c),(e)and(g)are as-cast ingots,figure(b),(d),(f) and (h) are heat-treated)

The isothermal oxidation kinetics curves of the four alloys at 1073K in laboratory air are showed in **Fig.1**. The gained weights of 0Cr0Mo, 2Cr0Mo, 0Cr2Mo and 2Cr2Mo alloys are 0.52684 mg/cm^2, 0.63218 mg/cm^2, 0.76357 mg/cm^2 and 0.64298 mg/cm^2 respectively, after the 200-hour oxidation test.The mass gain results of the alloys are much lower than that of normal TiAl binary alloy, which is about 1 mg/cm^2 or more at similar oxidation condition[13]. It's clear that all the three alloys have superior oxidation resistance compared to normal TiAl based alloy. It can be draw from the kinetics curves that Cr and Mo both rised the final oxidation weight.The 0Cr0Mo alloy exhibits superior oxidation resistance than the other three alloys. The weight of 2Cr0Mo alloy increased fast in the first 50 hours($5.27\times10^{-3}mg/(cm^2.h)$) and slowed down afterwards($1.05\times10^{-3}mg/(cm^2.h)$) in the last 100h, the oxidizing rate even less than that of 0Cr2Mo alloy's($1.31\times10^{-3}mg/(cm^2.h)$), which transformation means a formation of good oxidation scale that can prevent oxygen from coming in. The situation of 2Cr2Mo alloy is almost the same with 0Cr2Mo alloy, which means Mo barely has good effects in the oxidation resistance of the alloys.

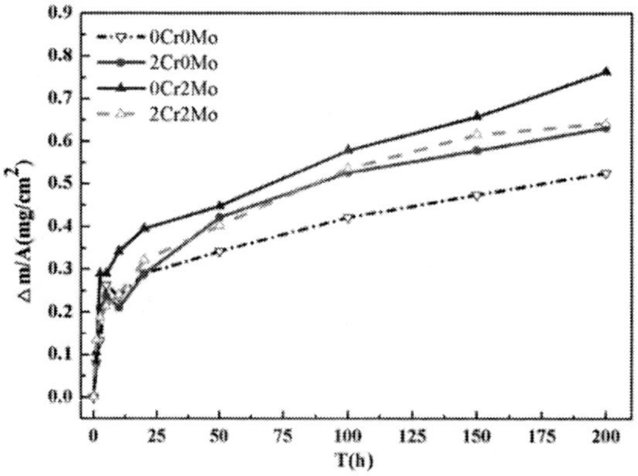

Figure 2. SEM Isothermal oxidation kinetics curves at 1073K of 0Cr0Mo, 2Cr0Mo, 0Cr2Mo and 2Cr2Mo alloy

Fig.3 shows the morphology of the oxidition scales of the four alloys after 200h oxidation test. **Fig.5** shows that the main oxides of the oxidation surface are Al_2O_3, TiO_2, Ti_3O, TiN, $(Al_{0.9}Cr_{0.1})_2O_3$ and SiO_2. There are only a very small amount of Mo oxide in the 0Cr2Mo and 2Cr2Mo alloy. **Fig. 3(a)** shows that 0Cr0Mo alloy has the weensiest structure in the three samples.There are barely any oxidation structure that sticks out, which surface morphology can keep steady oxidation resistance without any desquamating of the oxides. With the 2% at addition of Cr shown in **Fig 3(b)**, some sparse and scatter structure began to form on the surface, which differs from 0Cr0Mo alloy and grows clearly with Cr addition. **Fig.4 (a)** and **Fig.4 (b)** shows that the oxidition scale of 2Cr0Mo alloy is mainly two layers , in which the outer layer is much thicker than the inner.We can figure out from **Fig.6** that the inner layer are composed of titanic oxides and aluminium oxides which is an incompact oxidation structure , while the outer layer is almost titanic oxides which is poorer in oxidation resistance than Al_2O_3.

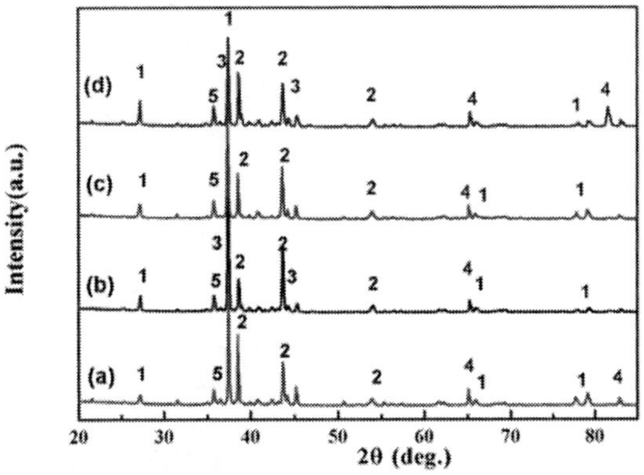

Figure 3. SEM scanning of the oxidized surface of (a)0Cr0Mo, (b)2Cr0Mo, (c)0Cr2Mo and (d)2Cr2Mo

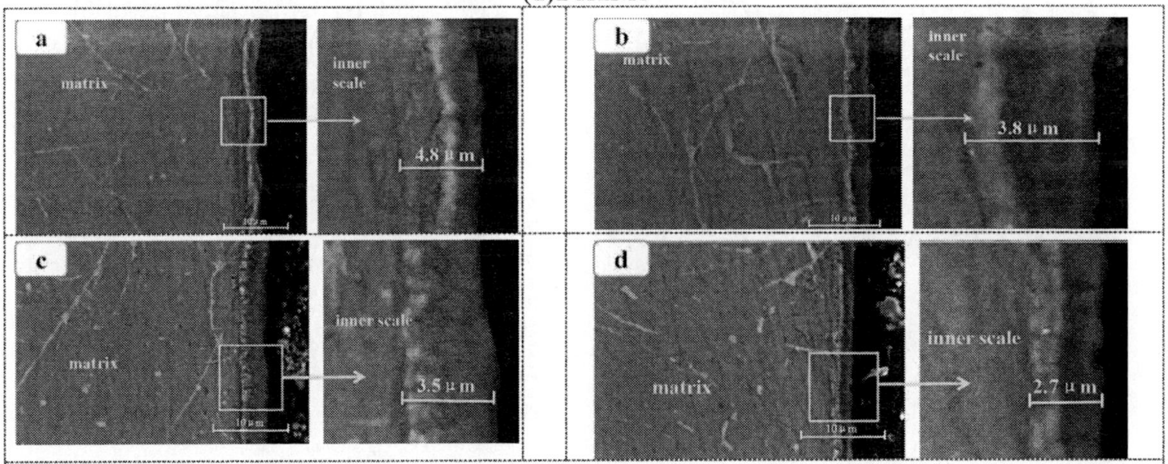

Figure 4. SEM scanning of oxidized section of (a)0Cr0Mo, (b)2Cr0Mo, (c)0Cr2Mo and (d)2Cr2Mo

The 0Cr0Mo alloy is mainly composed of three layers, in which the inner layer is the thickest in the three.We can find out in **Fig.6** that inner layer is titanic oxides and aluminium oxide mixture, middle layer is mostly titanic oxides, while outer layer is almost aluminium oxide.This kind of oxidation scale can keep most oxygen off the matrix from further oxidition. **Fig.3(c)** shows the oxidation surface of 0Cr2Mo alloy consists of tiny oxides with may gaps, which structure fails to hold back oxygen from the matrix of the ingot according to the oxidation kinetics curves.As can be learnt from the **Fig.4**,the thickness of the oxidition scale is of no responsible for the oxidation resistance. **Fig.6** indicates that the oxides of 0Cr2Mo alloy are mostly titanic oxides, while Mo barely have any positive effects on the oxidation resistance. As there are little Al_2O_3 in the oxidation scale, the increase weight of oxidation keeps increasing at an almost fixed rate. When Cr and Mo are added at the same time, the oxidation process of the 2Cr2Mo alloy is similer to the 2Cr0Mo alloy on the basis of the kinetics curves. However the **Fig.3(b)** and **Fig.3(d)** shows that the oxidition structure of 2Cr2Mo alloy has more scattered structures that protruded out than that of the 2Cr0Mo alloy. The structure may have more chances of peeling off and thus leads to further oxidation. The oxides in 2Cr2Mo alloy conclude more

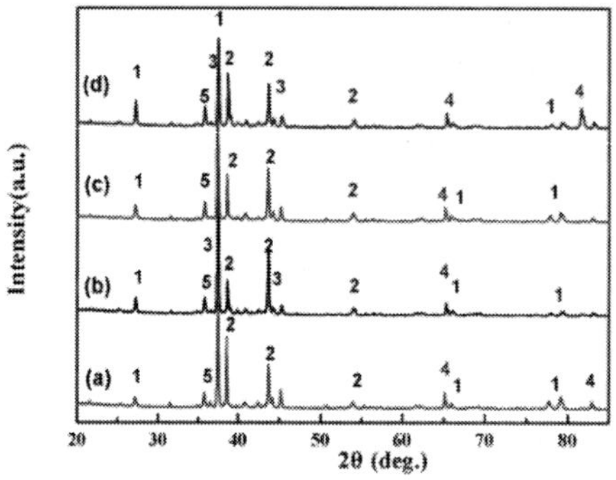

Figure.5. SEM scanning of the oxidized surface of (a)0Cr0Mo, (b)2Cr0Mo, (c)0Cr2Mo and (d)2Cr2Mo

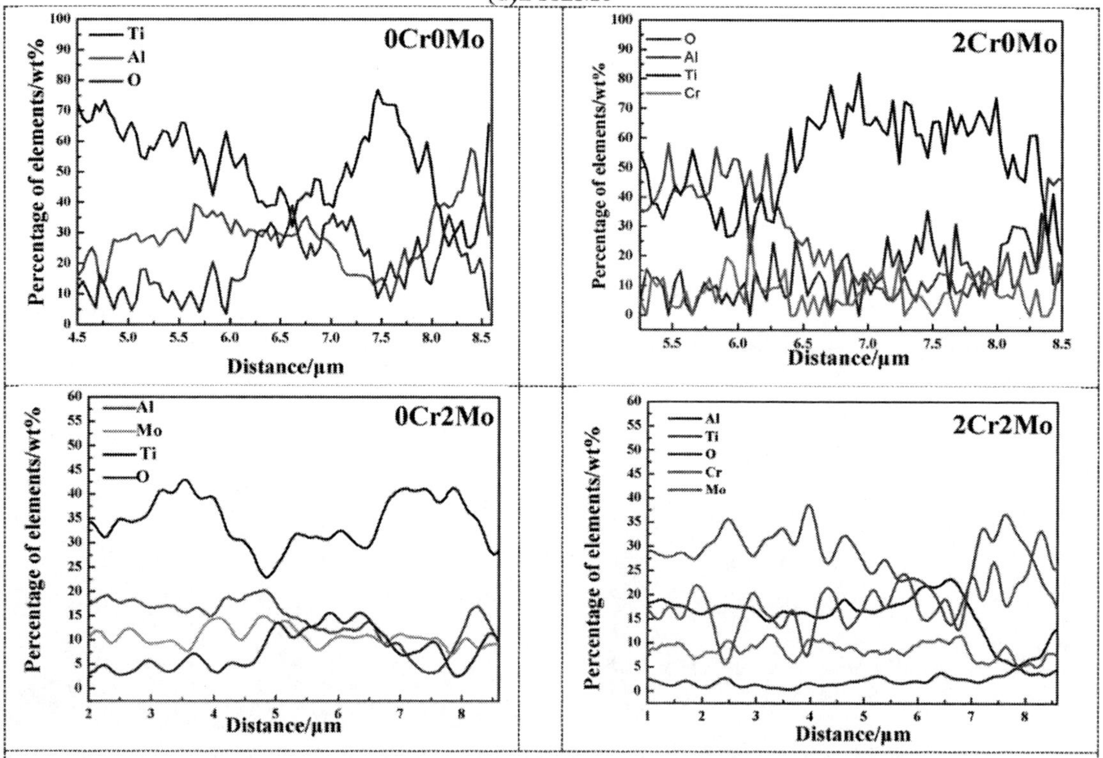

Figure 6. EDS analysis of main oxides in oxidation section of 0Cr0Mo, 2Cr0Mo, 0Cr2Mo and 2Cr2Mo

Cr and Al oxides that formed a $(Al_{0.9}Cr_{0.1})_2O_3$ oxide combination which is compact and can be a stopper of oxygen entrance. The adding of Cr may have caused the increase of the mass gain of the alloy, while it can really produce a compact oxidation structure that protect the matrix alloy from further oxidation. The adding of Mo alloy is not good for the alloy's oxidation resistance. The adding of both Cr and Mo has a similar oxidation rate with 2Cr0Mo alloy, but the adding of Mo still hinders the improvement of oxidation.

4. Conclusions

Four high Nb-TiAl alloys containing different percentages of Cr and Mo, i.e. 0Cr0Mo, 2Cr0Mo, 0Cr2Mo and 2Cr2Mo alloy were prepared. The conclusions can be drawn as follows:
(1) Cr and Mo modified alloy can both refine the microstructure by decreasing the grain size of the alloy and a transformation from fully lamellar structure to near fully lamellar structure.
(2)The heat treatment has a different effect on Cr and Mo modified alloys. Heat-treatment is more effective on Cr modified alloy than Mo modified alloy in driving the solid-state phase transformation from α_2 to γ
(3)The four alloys show superoir oxidation resistance over normal TiAl-based alloy. The adding of Cr incrased the oxidized mass gain of the alloy, but the $(Al_{0.9}Cr_{0.1})_2O_3$ oxide structure slows down the oxidation rate in the last 100h by filling the entrance of oxygen among the imcompact TiO_2/ Al_2O_3 oxidation structure. The adding of Mo alloy weakens the oxidation resistance of the alloy.
(4) The oxidation behavior of Cr and Mo modified alloy is similar to Cr modified alloy, the oxidation surface is different in flateness which matters a lot in long-term oxidation.

References

[1] Krause D, Lerch B, Locci I E. Development and evaluation of TiAl sheet structures for hypersonic applications[J]. *Materials Science & Engineering A*, 2007, 464(1):330-342.
[2] Dimiduk D M. Gamma titanium aluminide alloys—an assessment within the competition of aerospace structural materials[J]. *Materials Science & Engineering A*, 1999, volume 263(2):281-288(8).
[3] Shida Y, Anada H. The influence of ternary element addition on the oxidation behaviour of TiAl intermetallic compound in high temperature air[J]. *Corrosion Science*, 1993, (5-8):945-953.
[4] Taniguchi S, Uesaki K, Zhu Y C, et al. Influence of niobium ion implantation on the oxidation behaviour of TiAl under thermal cycle conditions[J]. *Materials Science & Engineering A*, 1998, 249(1):223–232.
[5] Liu Z C, Lin J P, Li S J, etal. Effects of Nb and Al on the microstructures and mechanical properties of high Nb containing TiAl base alloys[J]. *Intermetallics*, 2002, 10(7):653–659.
[6] He X, Yu Z, Lai X. Analysis of high temperature deformation behavior of a high Nb containing TiAl based alloy[J]. *Materials Letters*, 2008, 62(26):4181–4183.
[7] [7] Bystrzanowski S, Bartels A, Clemens H, et al. Creep behaviour and related high temperature microstructural stability of Ti–46Al–9Nb sheet material[J]. *Intermetallics*, 2005, 13(5):515–524.
[8] Chen G, Sun Z, Zhou X. Oxidation of Intermetallic Alloys in Ti-Al-Nb Ternary System[J]. *Corrosion*, 1992, 48(11):939-946.
[9] ZHAO Lili, LIN Junpin, WANG Yanli, etal. Early oxidation behaviors of Ti50Al and Ti45A18Nb alloys at high temperature[J]. *Acta Metallurgica Sinica*, 2008.
[10] CHEN G. L, ZHANG L. C. Deformation mechanism at large strains in a high-Nb-containing TiAL at room temperature[J]. *Materials Science & Engineering A Structural Materials Properties Microstructure & Processing*, 2002, 329(1):163–170.
[11] Kesler M S, Goyel S, Rios O, et al. A study of phase transformation in a TiAlNb alloy and the effect of Cr addition[J]. *Materials Science & Engineering A*, 2010, 527(12):2857-2863.
[12] Huang S C, Hall E L. The effects of Cr additions to binary TiAl-base alloys[J]. *Metallurgical & Materials Transactions A*, 1991, 22(11):2619-2627.
[13] Lin J P, Xu X J, Wang Y L, et al. High temperature deformation behaviors of a high Nb containing TiAl alloy[J]. *Intermetallics*, 2007, 15:668–674.
[14] Cheng T T, Willis M R, Jones I P. Effects of major alloying additions on the microstructure and mechanical properties of γ-TiAl[J]. *Intermetallics*, 1999, 7(1):89-99.
[15] Tian W H, Nemoto M. Effect of carbon addition on the microstructures and mechanical properties of γ-TiAl alloys[J]. *Intermetallics*, 1997, 5(3):237-244.
[16] Niu H Z, Chen Y Y, Xiao S L, et al. Microstructure evolution and mechanical properties of a novel beta γ-TiAl alloy[J]. *Intermetallics*, 2012, 31:225-231.

Acknowledgement This paper is funded by the International Exchange Program of Harbin Engineering University for Innovation-oriented Talents Cultivation.

IOP Conf. Series: Materials Science and Engineering **205** (2017) 012004 doi:10.1088/1757-899X/205/1/012004

The Effects of Different Electrode Types for Obtaining Surface Machining Shape on Shape Memory Alloy Using Electrochemical Machining

S G Choi[1], S H Kim[1], W K Choi[1], G C Moon[1], E S Lee[2,*]

[1]School of Mechanical Engineering, Inha Univ., 253 Yonghyun-Dong, Nam-Gu, Incheon, Korea.
[2]Department of Mechanical Engineering, Inha Univ., 253, Yonghuyn-Dong, Nam-Gu, Incheon, Korea.

Abstract. Shape memory alloy (SMA) is important material used for the medicine and aerospace industry due to its characteristics called the shape memory effect, which involves the recovery of deformed alloy to its original state through the application of temperature or stress. Consumers in modern society demand stability in parts. Electrochemical machining is one of the methods for obtained these stabilities in parts requirements. These parts of shape memory alloy require fine patterns in some applications. In order to machine a fine pattern, the electrochemical machining method is suitable. For precision electrochemical machining using different shape electrodes, the current density should be controlled precisely. And electrode shape is required for precise electrochemical machining. It is possible to obtain precise square holes on the SMA if the insulation layer controlled the unnecessary current between electrode and workpiece. If it is adjusting the unnecessary current to obtain the desired shape, it will be a great contribution to the medical industry and the aerospace industry. It is possible to process a desired shape to the shape memory alloy by micro controlling the unnecessary current. In case of the square electrode without insulation layer, it derives inexact square holes due to the unnecessary current. The results using the insulated electrode in only side show precise square holes. The removal rate improved in case of insulated electrode than others because insulation layer concentrate the applied current to the machining zone.

1. Introduction

Shape memory alloy (SMA) is applied to the various industries like medical industry and aero- space industry. Shape memory alloy has a special characteristic which is called the shape memory effect. Shape memory effect refers to the recovery of a deformed alloy to its original state through the application of temperature or stress to the alloy. For this reason, it is required to fabricate the micro and complex shape on SMA [1-2]. As this method of process that electrochemical machining, it can be machined on minute space of micro parts without stress and transmutation of material. The reason of achieving these advantages is non-contact machining between tool and workpiece, using metal dissolution by electrochemical reaction. Non-traditional machining methods, such as laser machining, electro-discharge machining, MR fluid machining, etching and Electro Chemical Machining, have been studied for smart materials [3-4]. Insulated electrode is required to acquire the precise results in electrochemical machining. The influence of the insulation layers which could affect current distribution has been investigated. In this paper, the square shape machining area were fabricated with electrochemical machining using the square electrode, the square electrode insulated only on the side. It is possible to obtain precise square shapes on the SMA using four type electrodes.

IOP Conf. Series: Materials Science and Engineering **205** (2017) 012004 doi:10.1088/1757-899X/205/1/012004

2. Experimental Setup

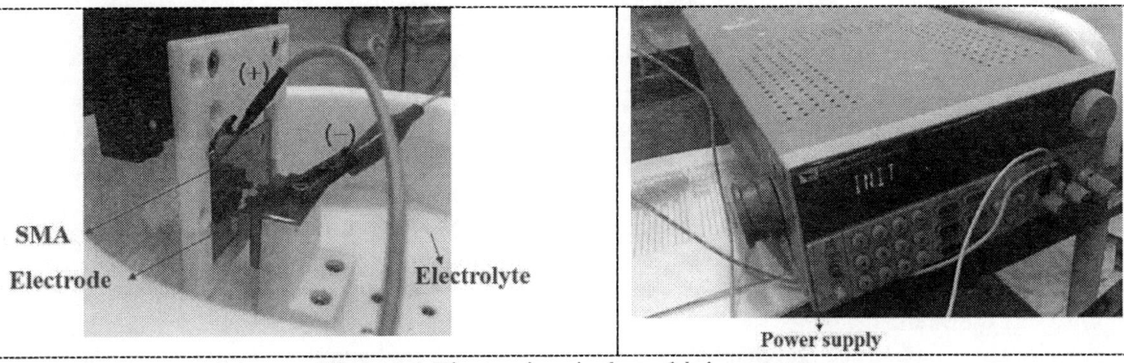

Figure. 1 Electrochemical machining system

Figure 1 shows the experimental electrochemical machining system. The workpiece was SMA with 51% Ni and 49% Ti. A cathode made of steel was used as the electrode. The anode was connected to the SMA, which was connected to a dedicated jig. A power supply supplies the current. Table. 2 shows the experiment conditions. Machining was performed by immersion in an electrolyte with an electrode gap of 1 mm. The constant current is employed in this experiment, and the current value is 30A. The appropriate electrolyte was determined in previous research, and it is found that NaCl electrolyte with a concentration of 0.27 wt% produces better workability than other electrolytes. Figure 2 shows the electrodes used in the experiment. The electrode was the general electrode, second electrode was the fillet, third electrode was non-insulated and finally electrode was the insulated.

Table 1. Experiment condition.

Conditions	Value
Current	30A
Electrolyte	NaCl 0.27 wt%
Electrode	General electrode, Fillet electrode, Non-insulated electrode, Insulated electrode
Electrode gap	1 mm
Workpiece	Shape memory alloy (SMA) (Ni: 51%, Ti: 49%)

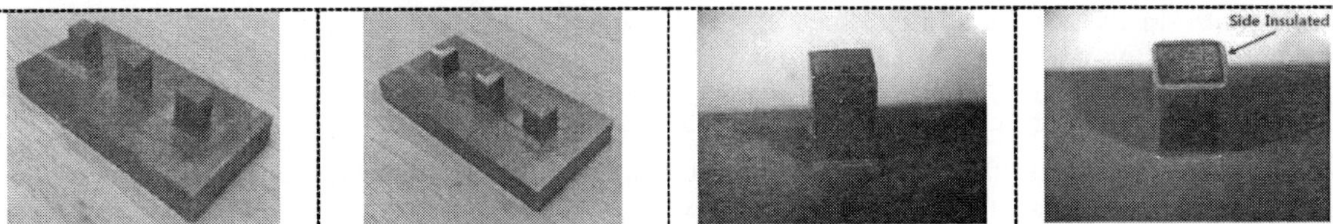

(a)General electrode | (b)Fillet of edge point | (c)non-insulated | (d)side of insulated

Figure. 2 Four type electrodes

3. Results and discussion

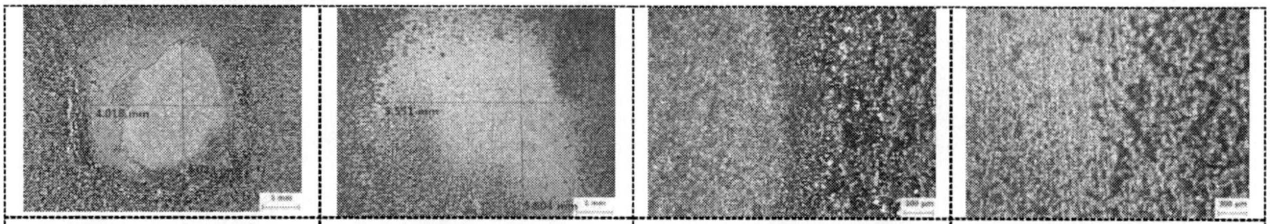

(a)General (x40) | (b)Fillet (x40) | (c)General (x100) | (d)Fillet (x100)

Figure. 3 Micrographics of first SMA experiment

Figure 3 shows micrographs of first SMA machined. Figure 3-(a), (b) shows a magnification of x40. Figure 3-(a) shows the machined shape from the general electrode. It was processed into a faint square shape. The size of electrode and machining shape size are different and the non-machining zones was also machined. Because a current flow in the non-machining zones. Figure 3-(b) shows the machining zones was expanded. In addition, the machining shape was not a square shape. Since the current flow due to the fillet electrode, current was spread to the outside. Figure 3-(c), (d) shows a magnification of x100. Also the left was machining zones and the right was non-machining zones. Since the concentrated of current density in machining zones, the surface is smooth. But, due to the imperfect current density, the machining surface was rough in the non-machining zones. The boundary line was clear in both zones. in case of the figure 3-(d), the boundary line was indistinct in the machining zones and non-machining zones. When the fillet edge, the direction of the current is spread out, it is affecting even by non-machining zones.

(a)not insulated (x40) | (b)insulated (x40) | (c)non-insulated (x100) | (d)insulated (x100)

Figure. 4 Micrographics of second SMA experiment

Figure 4 shows micrographs of second SMA machined. Figure 4-(a), (b) shows a magnification of x40. Blue circle was the machining zone, machining lines was the border zone, the other zone was the non-machining zone. Figure 4-(a) shows the machining shape using not insulated electrode. The non-machining zones as well as the machining zone was machined. Although the machining shape was a square shape, do not appear correctly and the machining lines are irregular, the machining surface was not smooth. In case of the non-insulated electrode without insulation layer, it derives inexact square shapes due to the unnecessary current. Figure 4-(b) shows the machining shape using side insulated electrode. The non-machining zones was not machining and the current was concentrated in the machining zones has been machining. The machining shape was shown clearly in a square shape. In addition, the machining lines are regularly and machined surface has been processed uniformly. The results using the insulated electrode show precise square shapes. Figure 4-(c), (d) shows a magnification of x100. The machining surface was not smooth in the machining zone. Non-machining zone was also machining, because of not insulated electrode, and due to excessive current, the border zone was also irregular. The machining surface was very smooth in the machining zones and the non-machining zone was not machined. Due to excessive current, the interface was also irregular. The border zone was a little regularly because the current concentrated in the machining zones.

4. Conclusion

Through this study, the analysis in machining shapes of shape memory alloy (SMA) according to four electrodes. First experiment using the general electrode and fillet electrode. In the case of the fillet electrode, the appearance was emitted small, the current is concentrated in the fillet point. And the current was spread to the outside. If the fillet of edge point was the current is spread widely. In case of the non-insulated electrode without insulation layer, it non-machining zone as well as the machining zone was machined. When the fillet edge, the direction of the current is spread out, it is affecting even by non-machining zones.
Second experiment using the non-insulated electrode and insulated electrode. The results using the insulated electrode in only side show precise square shapes. The most precise square shapes have been found when the square electrode insulated is applied because the insulation layer concentrate the unnecessary current to machining zone. The removal rate improved in case of square electrode insulated than others because insulation layer concentrate the applied current to the machining zone.

5. References

[1] R. Mukherjee, T. F. Christian and R. A. Thiel, An actuation system for control multiple shape memory alloy actuator, Sensor and Actuators A:physics, 55 (1996) 185-192.
[2] M. Manjaiah, S. Narendranath, S. Basavarajappa, Review on non-conventional machining of shape memory alloys, Transactions of Nonferrous Metals Society of China 24(2014) 12-21.
[3] M. J. Shin, S. Y. Baek and E. S. Lee, A study for improving surface roughness and micro-deburring effect of nitinol shape memory alloy by electropolishing, Transactions of the Korean Society of Machine Tool Engineers,16 (6) (2007) 49-54.
[4] C. Xuezhen, X. Zhengyang, Z. Dong, F. Zhongding, Z. Di, Experimental research on electrochemical machining of titanium alloy Ti60 for a blisk, Chinese Journal of Aeronautics (2015).

Acknowledgments
This research was supported by the Strengthening competitiveness business of industrial clusters funded by the Korea Industrial Complex Corporation.

IOP Conf. Series: Materials Science and Engineering **205** (2017) 012005 doi:10.1088/1757-899X/205/1/012005

Lap Joints of 6061 Al Alloys by Friction Stir Welding

M. A. Tashkandi[1]

[1]Department of Mechanical Engineering, College of Engineering, Northern Border University. P.O. Box 1321, Arar 91431, Kingdom of Saudi Arabia
tashkandi@gmail.com; mtashkandi@nbu.edu.sa

Abstract. Lap joints of 6061 aluminium alloys were obtained using friction stir welding. The welding parameters used are: rotational speed, welding speed and pin Length. There were two levels of each factor involved. Rotational speed levels were 900 and 1120 rpm, welding speed levels were 1 and 1.5 mm/sec and pin length levels were 4.5 and 5.5 mm. Successful lap joints were produced for the different combinations of the above welding parameters. The joints quality tested under tensile testing and the microstructure of the welding zone were studied. Fracture mostly occurred in the base metal for most lap-shear welded joints. The exception was joints with welding parameters as follows: rotational speed 900 rpm, the welding speed 1.5 mm/sec and pin length 4.5 mm. In this case, the fracture occurred in the welding zone.

1. Introduction

Friction stir welding (FSW) is a solid-state joining process. It was invented at The Welding Institute (TWI) of UK in 1991 as a solid-state joining technique, and it was initially applied to aluminum alloys [1–3]. Lap joints are widely used in the assembly of parts and products in the transportation industry. Common examples include ship decks, railway tankers, goods wagons, and stringer to skin in aircraft fuselages. Most aircraft and aerospace sheet metal structures involve lap joints over a range of different material joint combinations [4, 5], so there has been growing interest in recent years in utilizing FSW for joining Al alloys and Al-Cu alloys [1] in lap configuration. Conventionally, these joints are produced using riveting, given the limited fusion weldability of many high-strength Al alloys and difficulties in welding Al-Cu alloys. This also because of the difference in physical and mechanical properties of both metals, as well as its chemical affinity at temperatures higher than 120°C [5, 6]. Thus using FSW instead of riveting for joining Al alloys and Al-Cu alloys in lap configuration can help achieve significant weight and cost savings with improved mechanical performance and reduced manufacturing complexity [7].

FSW of lap joints is obtained by fixing the adjoining blank edges on a properly designed fixture, inserting a specially designed rotating pin into the sheets to be welded, penetrating them both and then moving the tool all along the welding line. A proper support has to be given to the joining edges in order to avoid the forward extrusion of the blanks material due to the tool pin action. Furthermore, the upper sheet must be properly supported to prevent any bending mechanics during the welding process [8]. The friction caused by the rotating tool shoulder at the upper sheet surface heats the sheets up and a thermal softening of sheet materials occurs at a region under the tool shoulder. This allows the tool pin to stir up the sheets materials, mix them together and make the welded joint without any melting taking place [4, 9–11]. However, there is another strategy to produce FSW lap joints, in which the process is done by "pin-less" tool configuration and using filler material lying between the two sheets [12].

Recently many researchers investigated process parameters that significantly affect the quality of lap joints produced using different Al alloys such as AA 5083 [13], AA 2024 [14], and AA 7075 [11]. In general, the quality of lap joints produced by friction stir welding depends mainly on weld exists and extends in the form of three different defects: hooking defects, kissing bonds and top sheet thinning. The hooking defects can be seen mostly at the thermo-mechanical affected zone of the advancing side when the sheets interface is pulled up into the top sheet. This defect causes a local thinning at the top sheet and works as crack initiator which affects joint strength. The kissing bond defect results from insufficient heat transfer to the interface of the sheets, thus leading to separated interfaces with some resident oxide layers among them. This defect usually occurs at the retreating side because of its lower temperature during friction stir welding. Finally, top sheet thinning is observed on both sides, this thinning is due to the upward pulling of the interface into the top sheet thereby reducing its cross section [11].

Several attempts were carried out to fully understand FSW process mechanics [8] in order to overcome the above stated defects and consequently improve joint efficiency. Geometrical parameters and technological parameters are the main factors that affect joint efficiency. Geometrical parameters include: dimension of the shoulder, dimensions of the pin, and the pin profile. Technological parameters include: tool rotating speed, tool feed rate, and the tool depth sinking into the sheets to be welded. These process parameters have to be properly chosen on the basis of the blank material in order to maximize the joint strength [4, 7–9, 15].

In all the examined researches it was found that to improve the mechanical resistance of the obtained FSW lap joints, a significant consideration must be given to the geometrical characteristics of the tool; in particular, the diameter of the pin. Turns out that a large tool pin diameter is required in order to extend as much as possible over the welded surface. However, large tool shoulder diameter is also required in order to increase the heat flux generated by the frictional forces [4, 8].

Emad et al. [4] carried a friction lap stir weld for AA5456 sheets of two thicknesses 5 and 2.5mm. They studied the influence of the tool geometry and the rotation speed on the strength of the lap joints. They used four different pin geometries (conical thread pin, cylindrical-conical thread, stepped conical thread, and flared triflute) with two rotational speed 600 and 800rpm. It was found that the optimum lap joints properties were obtained using the stepped conical pin and the rotation speed of 600rpm. Babu et al. [7] used two different tools (triangular and threaded taper cylindrical pins) to produce friction stir lap welds in 3mm thick Alclad sheets of Al alloy 2014-T4. The effects of tool geometry on weld microstructure, lap-shear performance and failure mode were investigated. It was found that the pin profile can significantly influence the hook geometry of the weld. Thus affecting the joint strength and failure modes. Buffa et al. [8] reported on investigating lap joining AA2198-T4 by FSW through experimental and numerical analyses. The joint configuration, tool geometry and rotational speed were varied. It was found that using cylindrical-conical pin tools and the appropriate relative sheet positioning increase the welded nugget extension and integrity. Thus, improving the mechanical performances of the obtained joints.

The effects of tool travel and rotation speed on weld quality and defect generation were also investigated. Bisadi et al. [10] investigated the effects of FSW process parameters including rotational and welding speeds on the microstructure and mechanical properties of aluminium 5083 alloy in lap joint welding and analysed different joint defects. They reported that the nugget area had the best grain size as well as higher hardness in comparison with other welding areas. It was also reported that the best joint properties were achieved at a rotational speed of 825rpm and a welding speed of 32mm/min.

There seem to be vigorous investigations on friction stir welding of different types of Al alloys but the studies that considered FSW on Al6061 alloys are very few. Moreover, all the cited work above were focused into tool geometry and/or process parameters; mainly tool rotational speed and welding speed. As a result, the current study will investigate FSW on Al6061 alloys considering the rotation speed of the tool, the welding speed and also the pin length to obtain a set of optimum conditions for welding Al6061 alloys through FSW.

2. Experimental Work

A 6061 aluminum alloy plates with a thickness of 3 mm were used to fabricate lap joints by friction stir welding process. The chemical composition of the studied material is given in table 1. Two rectangular aluminum plates of 300 x 150 mm were lap-welded by friction stir welding process. A conventional vertical milling machine was used for the welding process. The head of the milling machine was tilted 0.5 degrees and a special suitable fixture was designed to hold the plates during the welding process. The fixture was tightly secured on the milling machine table to prevent vibration occurring as a result of the frictional forces of the welding process. The lap joints were prepared for single lap-shear load.

Table 1: Chemical composition of Al6061 alloy

Si	Mg	Cu	Fe	Mn	Ti	Zn	Cr	Al
0.75	0.9	0.5	0.5	0.15	0.05	0.03	0.03	Rest

The welding tool was made out of high carbon steel with 24mm shoulder diameter and 20mm height. A 6mm square pin profile was used in this investigation. The welding parameters used in this investigation are given in table 2. Microstructure examinations were done using an optical microscope. The specimens, which were taken from the middle of the stirred zone, were prepared by standard metallographic techniques and etched with Keller's reagents to reveal the grain structure. A lap-shear fracture load are prepared according to ASTM: D 1002-01 for single lap shear fracture load. The tensile test specimens had a shape as shown in Fig.1. The test specimens are place in the grip of WDW -20 computer controlled electromechanical universal testing machine and pulled at 1.3 mm/min.

Table 2: The used welding parameters for single lap-shear load

Experiment no.	Rotational speed rpm)	Welding speed (mm/sec)	Pin length (mm)
1	900	1	5.5
2	900	1.5	5.5
3	900	1	4.5
4	900	1.5	4.5
5	1120	1	5.5
6	1120	1.5	5.5
7	1120	1	4.5
8	1120	1.5	4.5

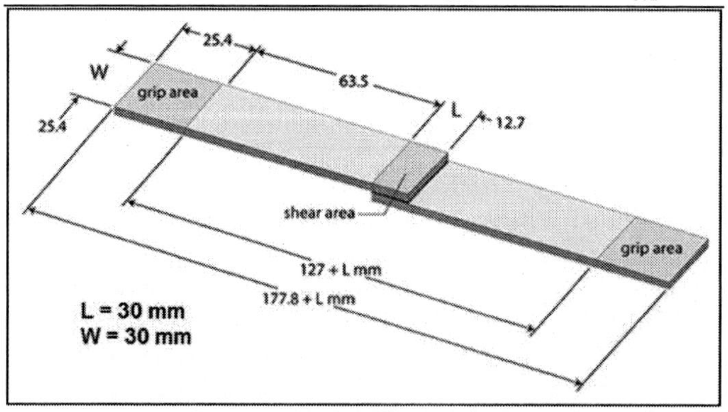

Figure. 1. Tensile specimen

Further analysis of the experiment was also carried out. A characterization of the adhesive bonds between the welded surfaces was necessary to further investigate the performed welding in the study; namely a peel test. Peel testing is used to determine the maximum peel load extensively to evaluate the bonding strength of tape, adhesives and flexible substrates, including rubber, films, biomaterials, dental materials, medical packaging and consumables. This method is used to test lap welded joints within the standard parameters recommended by ASTM. In this study four samples are peel tested to study the effect of welding parameters on the peak peel load.

3. Result and Discussion

3.1. Microstructure

Microstructure of 6061 aluminium alloy lap joint produced by friction stir welding and the base metal was studied by employing optical microscopy. In this study the microstructure of the stir zone only was studied. Fig. 2 shows a typical microstructure of the base metal used in these experiments. Heat input during welding process controls the temperature of welding zone. The heat input during the welding process depends on the friction stir welding parameters. Heat input increased as the rotational speed increased and is also proportional to applied axial force and the contact area between shoulder face and the welded surface. As the rotational speed increases, the heat input during the friction stir welding process also increases. The evolution of microstructure in the stir zone (SZ) depends on the recrystallization temperature and the cooling rate of welding region. The stir zone of welded lap plates was cooled by ambient air.

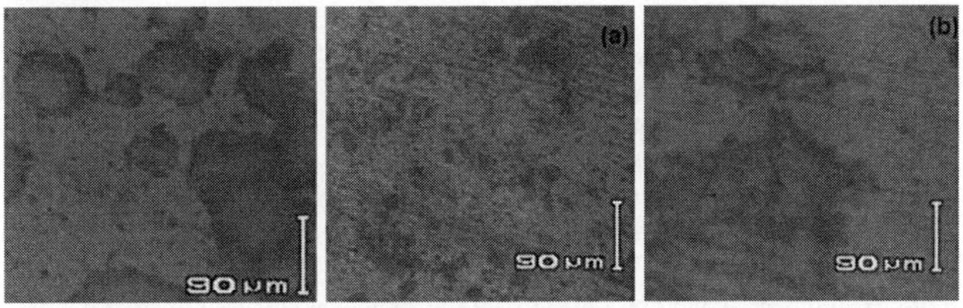

Figure. 2. Microstructure of the base material on the left. Microstructures of the welding joints welded (a) 900 rpm, and (b) 1120 rpm

The effect of the tool rotational speed on the microstructure of the stir zone was considered. The welding speed and the pin length were maintained at 1.5 mm/sec and 5.5 mm respectively whereas the rotational speed changed from 900rpm in the first case to 1120 rpm in the second case. Fig. 3 shows the effect of the rotational speed on the microstructure of the stir zone. At a rotational speed of 900 rpm, the SZ zone is composed of a fine equiaxed grains formed under high temperature and large deformation in the weld centre due to the stirring process (Fig. 2a). At higher rotational speeds (1120 rpm), more heat is induced during the welding process which gives prolonged heat dissipation with constant welding speed. Fig. 2b shows coarser and deformed grain size. Thus, that the heat input during welding process controls the temperature in the welding area.

The effect of the welding speed is considered next. Fig. 3 shows the effect of the welding speed on the microstructure of the stir zone. At low welding speeds (1 mm/s), the microstructure has a deformed and coarser grain size where observed, Fig. 3a. This is due to the additional heat input that leads to severe plasticity of the metals under the shoulder face. At higher welding speeds (1.5 mm/sec), the microstructure shows deformed and a coarse grain size but with less deformation and size as compared to the microstructure when the welding speed was 1mm/sec, Fig. 3b. Thus, the welding speed controls the rate of heat input during welding per unit length.

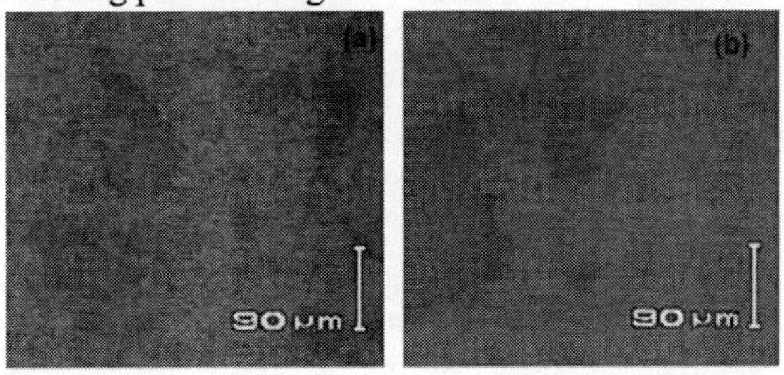

Figure. 3. Microstructure of welding joints welded at pin length of 5.5mm, rotational speed of 1120 rpm and a welding speed of (a) 1mm/sec and (b) 1.5mm/sec.

IOP Conf. Series: Materials Science and Engineering **205** (2017) 012005 doi:10.1088/1757-899X/205/1/012005

3.2. Mechanical Properties

The strength of a lap-welded sample cannot be expressed using the normal load/area, as the stress distribution along the joint area during tensile-shear test is highly uneven [16]. Instead, lap-shear fracture load is used in this study to express joint strength. The quality of the produced lap joints by friction stir welding depends mainly on exists. Which, may extend of three different defects: top sheet thinning, kissing bonds and hooking defects [11]. However, the existence of these defects is commonly influenced by many parameters. The most effective parameters in friction stir welded joints are the tool characteristics and process the parameters such as the rotational and welding speeds the process [17].

In this study, the effect of rotational speed, welding speed, and pin length on the lap-shear fracture load of 6061 aluminium lap welded joints was investigated. Table 3 shows welding parameter combinations and the lap-shear fracture load. In addition, it also shows the fracture location of the tested joints observed in this study. It is clearly shown that all of the tested samples are fractured outside the stir zone except sample 4 which fractured within the stir zone. This sample was welded at a rotational speed of 900r pm, welding speed of 1.5 mm/sec, and pin length of 4.5 mm. The cross sectional area of the welding area is twice the area of the base metal in lap-joints. Also, the thickness of the welding area is twice the thickness of the base metal since both plates have the same thickness. This indicates that good bonding resulted from successful mixing will occur in the welds. As a result, weldments are most probable to fracture under tensile testing in the base metal particularly in the heat affected zone and/or the thermos-mechanically affected zone.

Sample 5 fractured at the heat affected zone at the top sheet in advanced side due to sheet thinning unlike sample 4 that was discussed earlier. Sample 4 had a pin length of 4.5 mm, and a rotational speed of 900 rpm with a high welding speed of 1.5 mm/sec. This low rotational-high welding speeds combination allowed less heat per unit length to be induced to the welding area. Low welding temperature may produce insufficiently stirred sheet interface and cause many defects like kissing bonds. On the other hand, extremely high welding temperature may lead to unexpected joint microstructure or alloy melting.

Table 3: Strength and elongation of the samples tested under tensile test

Sample 1		**Sample 2**		**Sample 3**	
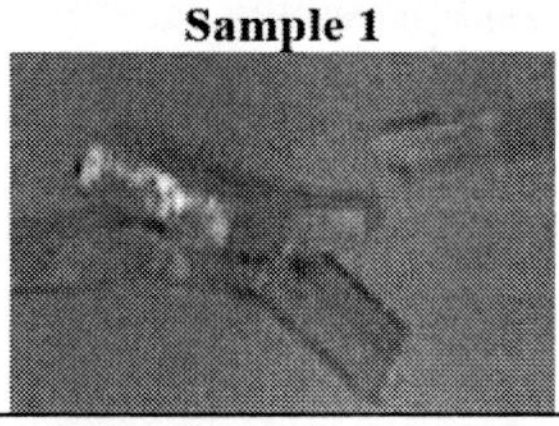		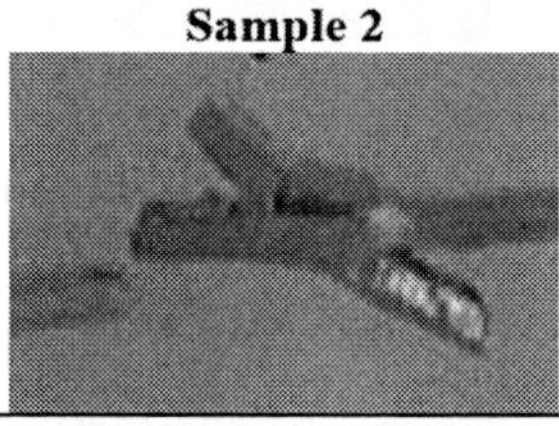		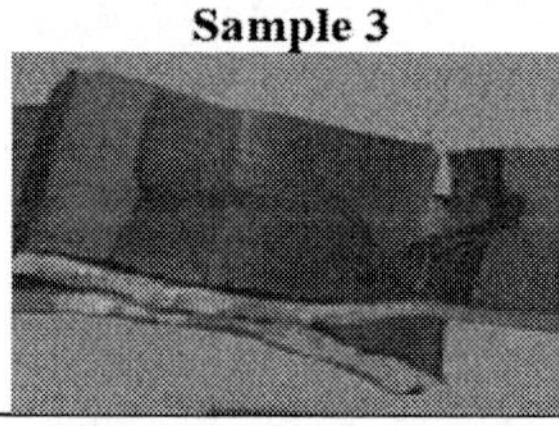	
Rotational speed	900 rpm	Rotational speed	900 rpm	Rotational speed	900 rpm
Welding speed	1 mm/sec	Welding speed	1.5 mm/sec	Welding speed	1 mm/sec
Pin length	5.5 mm	Pin length	5.5 mm	Pin length	4.5 mm
Elongation	13.71mm	Elongation	10.16 mm	Elongation	9.88 mm
Fracture load	5908 N	Fracture load	5796 N	Fracture load	4780 N

Sample 4		**Sample 5**		**Sample 6**	
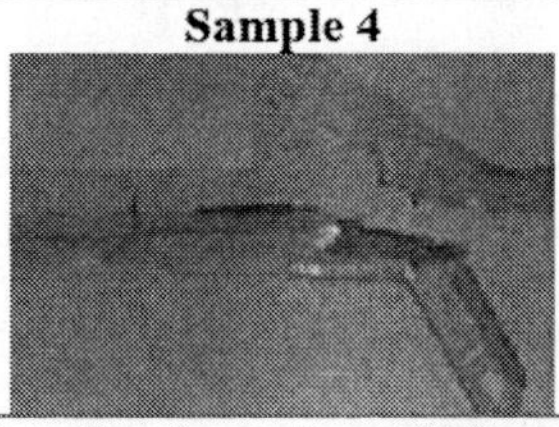		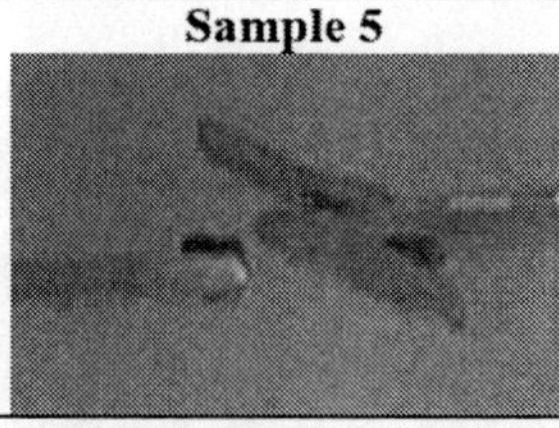		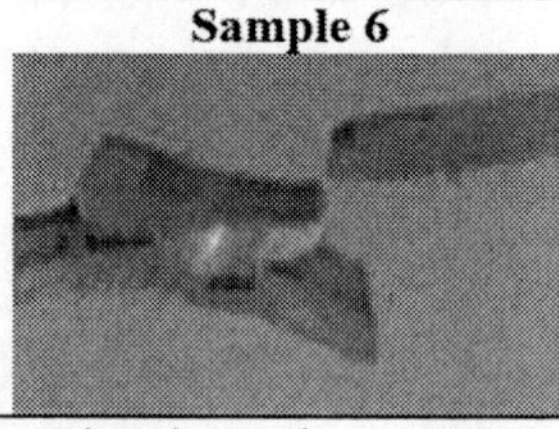	
Rotational speed	900 rpm	Rotational speed	1120 rpm	Rotational speed	1120 rpm
Welding speed	1.5 mm/sec	Welding speed	1 mm/sec	Welding speed	1.5 mm/sec
Pin length	4.5 mm	Pin length	5.5 mm	Pin length	5.5 mm
Elongation	6.56 mm	Elongation	10.86 mm	Elongation	10.27 mm
Fracture load	4469 N	Fracture load	5932 N	Fracture load	6092 N

Sample 7

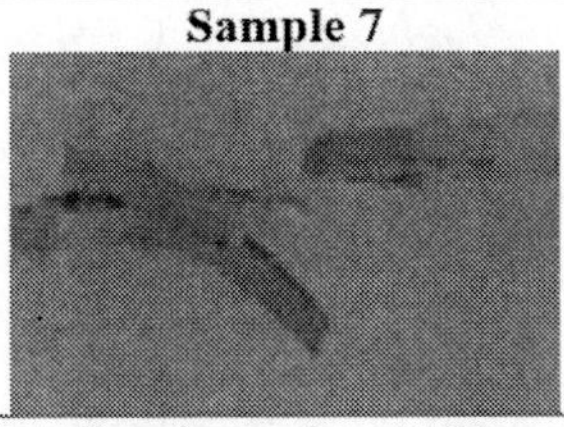

Sample 8

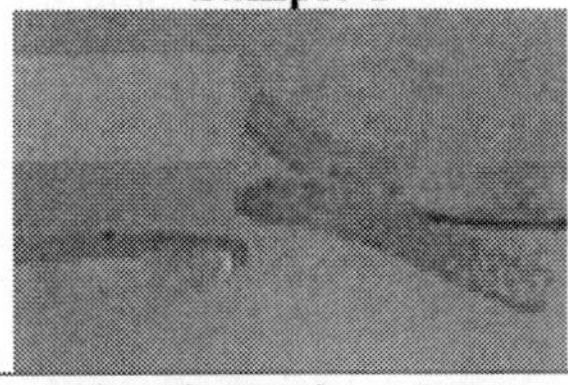

Sample 7		Sample 8	
Rotational speed	1120 rpm	Rotational speed	1120 rpm
Welding speed	1 mm/sec	Welding speed	1.5 mm/sec
Pin length	4.5 mm	Pin length	4.5 mm
Elongation	9.33 mm	Elongation	11.09 mm
Fracture load	5872 N	Fracture load	5952 N

3.3. Peel Test

It has been found that the joint welded at a rotational speed of 1120 rpm and welding speed of 1.5 mm/s possess the highest peak peel load as given table 4 sample 2. This result due to suitable balance between rate heat generated due to high rotational speed with the rate of heat dissipation that happened by the forward moving of the tool. Sample 4 in table 4 possess the lowest peak peel load where the joint produced under a rotational speed of 900 rpm and welding speed of 1 mm/s. Fig. 4 shows the fractured samples under peel test. It has been shown that samples 1, 2 and 3 were resist the fracture where some elongation happened to the samples before fracture, where it clear in the samples legs where some cracks appeared due to elongation and reduction in area.

Table 4: Peak peel load of the tested samples under peel test

Sample No.	Rotational speed rpm	Welding speed mm/s	Pin length mm	Fracture Load, N	Location of fracture
1	1120	1	5.5	1672	Welding area
2	1120	1.5	5.5	3024	
3	900	1.5	5.5	2304	
4	900	1	5.5	672	

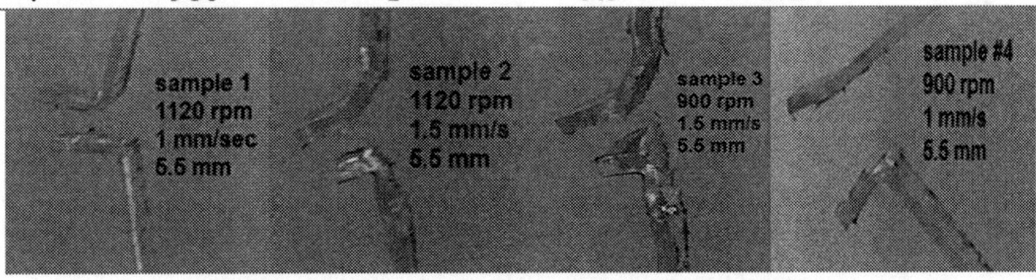

Figure. 4. Peel fracture of different samples

4. Conclusion

In this research, friction stir welding of Al6061 alloys was successfully accomplished. Sheet metals were successfully lap-joint welded for a specific range of process parameters. The tool rotational speed seemed to affect the microstructure of the welded material; as the speed increased from 900 to 1120 rpm, the microstructure became coarser and showed more deformation. In addition, high quality lap welded joints can be achieved by adjusting the welding parameters so that a balance between heat input and heat dissipation is reached. As far as the lap-shear fracture load, the welded joints in this investigation seemed to have relatively high lap-shear fracture load. Moreover, the fracture in most cases was observed in the base metal instead of the joint itself. Finally, the lap-joints produced at a rotational speed of 1120 rpm and a welding speed of 1.5 mm/sec had the highest fracture load and seemed to fail at the heat affected zone.

References

[1] D. Baffari, G. Buffa, D. Campanella, L. Fratini, and F. Micari, "Friction based Solid State Welding Techniques for Transportation Industry Applications," *Proceedings of the International*

Conference on Manufacturing of Lightweight Components - ManuLight 2014, vol. 18, pp. 162–167, 2014.

[2] R. M. Leal, C. Leitão, A. Loureiro, D. M. Rodrigues, and P. Vilaça, "Material flow in heterogeneous friction stir welding of thin aluminium sheets: Effect of shoulder geometry," *International Conference on Recent Advances in Composite Materials (ICRACM 2007)*, vol. 498, no. 1–2, pp. 384–391, 2008.

[3] S. R. Ren, Z. Y. Ma, and L. Q. Chen, "Effect of welding parameters on tensile properties and fracture behavior of friction stir welded Al–Mg–Si alloy," *Scripta Materialia*, vol. 56, no. 1, pp. 69–72, 2007.

[4] E. Salari, M. Jahazi, A. Khodabandeh, and H. Ghasemi-Nanesa, "Influence of tool geometry and rotational speed on mechanical properties and defect formation in friction stir lap welded 5456 aluminum alloy sheets," *Materials & Design*, vol. 58, pp. 381–389, 2014.

[5] I. Galvão, D. Verdera, D. Gesto, A. Loureiro, and D. M. Rodrigues, "Influence of aluminium alloy type on dissimilar friction stir lap welding of aluminium to copper," *Journal of Materials Processing Technology*, vol. 213, no. 11, pp. 1920–1928, 2013.

[6] P. Xue, B. L. Xiao, D. R. Ni, and Z. Y. Ma, "Enhanced mechanical properties of friction stir welded dissimilar Al–Cu joint by intermetallic compounds," *Materials Science and Engineering: A*, vol. 527, no. 21–22, pp. 5723–5727, 2010.

[7] S. Babu, G. J. Ram, P. V. Venkitakrishnan, G. M. Reddy, and K. P. Rao, "Microstructure and Mechanical Properties of Friction Stir Lap Welded Aluminum Alloy AA2014," *Journal of Materials Science & Technology*, vol. 28, no. 5, pp. 414–426, 2012.

[8] G. Buffa, G. Campanile, L. Fratini, and A. Prisco, "Friction stir welding of lap joints: Influence of process parameters on the metallurgical and mechanical properties," *Materials Science and Engineering: A*, vol. 519, no. 1–2, pp. 19–26, 2009.

[9] J. Mohammadi, Y. Behnamian, A. Mostafaei, and A. P. Gerlich, "Tool geometry, rotation and travel speeds effects on the properties of dissimilar magnesium/aluminum friction stir welded lap joints," *Materials & Design*, vol. 75, pp. 95–112, 2015.

[10] H. Bisadi, A. Tavakoli, M. Tour Sangsaraki, and K. Tour Sangsaraki, "The influences of rotational and welding speeds on microstructures and mechanical properties of friction stir welded Al5083 and commercially pure copper sheets lap joints," *Materials & Design*, vol. 43, pp. 80–88, 2013.

[11] L. Dubourg, A. Merati, and M. Jahazi, "Process optimisation and mechanical properties of friction stir lap welds of 7075-T6 stringers on 2024-T3 skin," *Materials & Design*, vol. 31, no. 7, pp. 3324–3330, 2010.

[12] B. Kuang, Y. Shen, W. Chen, X. Yao, H. Xu, J. Gao, and J. Zhang, "The dissimilar friction stir lap welding of 1A99 Al to pure Cu using Zn as filler metal with "pinless" tool configuration," *Materials & Design*, vol. 68, pp. 54–62, 2015.

[13] H. Bisadi, M. Tour, A. Tavakoli, "The Influence of Process Parameters on Microstructure and Mechanical Properties of Friction Stir Welded Al 5083 Alloy Lap Joint," *American Journal of Materials Science*, vol. 1, no. 2, pp. 93–97, http://journal.sapub.org/materials, 2011.

[14] D. Fersini and A. Pirondi, "Fatigue behaviour of Al2024-T3 friction stir welded lap joints," *Fracture and Damage in Engineering ApplicationsXXXIV National Conference of the AIAS*, vol. 74, no. 4, pp. 468–480, 2007.

[15] G.-F. Zhang, W. Su, J. Zhang, Z.-X. Wei, and J.-X. Zhang, "Effects of shoulder on interfacial bonding during friction stir lap welding of aluminum thin sheets using tool without pin," *Transactions of Nonferrous Metals Society of China*, vol. 20, no. 12, pp. 2223–2228, 2010.

[16] S. Y. Z.W. Chen, "Friction Stir Lap Welding: material flow, joint structure and strength," *Journal of Achievements in Materials and Manufacturing Engineering*, vol. 55, no. 2, pp. 629–637, 2012.

[17] M. Grujicic, G. Arakere, C.-F. Yen, and B.A. Cheesema, "Computational Investigation of Hardness Evolution During Friction-Stir Welding of AA5083 and AA2139 Aluminum Alloys," (English), *Journal of Materials Engineering and Performance*, vol. 20, no. 7, pp. 1097–1108, 2011.

IOP Conf. Series: Materials Science and Engineering **205** (2017) 012006 doi:10.1088/1757-899X/205/1/012006

The Effect of Sugarcane Bagasse`s Size on the Properties of Pretreatment and Enzymatic Hydrolysis

Xu Jun[1*, 2], Zhou Guoqiang[1], Li Jun[1]

[1]State Key Laboratory of Pulp and Paper Engineering, Plant micro/nano fiber research center, South China University of Technology, Guangzhou 510640, China
[2] Key Laboratory of Pulp and Paper Science & Technology of Ministry of Education of China, Qilu University of Technology, Jinan 250353, China

Corresponding author: xujun@scut.edu.cn

Abstract. The influence of milled bagasse particle size on their reducing sugar and lignin content during dilute acid hydrolysis followed by enzymolysis was investigated. The biomass crystal structures of hydrolyzed residues and enzymolyzed substrates were studied with X-ray diffractometry (XRD). The results showed that the conversion ratio of reducing sugar declined with decreasing milled bagasse particle size. The conversion ratio of reducing sugar after acid hydrolysis decreased from 31.3% to 28.9%. The smaller of the milled bagasse particle size was, the higher of the klason lignin content of hydrolyzed residuals was, which resulted in a decline in conversion ratio of reducing sugar during enzymolysis. In this study, the optimal size of milled bagasse particles was 10 to 20 meshes. The total reducing sugar conversion ratio was 61.5%, consisting of 31.3% in hydrolysis and 30.2% in enzymolysis. After hydrolysis, the specific surface area and pore size increased, and the fiber length was shortened. The inner microfiber bundles were exposed, which improved the accessibility of cellulase and the efficiency of enzymolysis.

1.Introduction

Plant fiber is an important renewable resource that is mainly composed of cellulose, hemicellulose, and lignin. Lignin and hemicellulose are connected by covalent bonds, and cellulose is encrusted with lignin and hemicellulose.[1] Due to its complex structure and the crystalline structure of cellulose, fibers are resistant to enzymatic hydrolysis.[2] However, pretreatment of lignocellulosic materials improves the sugar yield, which can also increase the specific surface area of cellulose.[3] Acid pretreatment is a common chemical pretreatment method.[4] Acid dissolves hemicellulose in wood fiber and improves the accessibility of cellulose.[5] Due to the highly concentrated acid, the equipment used for pretreatment must have high corrosion resistance. So, diluted acid with mild reaction conditions and less inhibitors are widely used.[6] Many factors influence the efficiency of cellulase, such as the size of milled bagasse particles, lignin and hemicellulose content, crystallinity, and the accessibility of cellulose.[7] The shape and size of milled bagasse particles determine the external specific surface area, and the porosity and capillary structure in fiber determine the internal specific surface area.[8] Efficient pretreatment methods effectively improve the internal and external surfaces of milled bagasse particles, especially the internal specific surface area. The external surface can also be improved by reducing the size of milled bagasse particles.[9]

In order to improve the ethanol conversion ratio during enzymatic hydrolysis and fermentation[10], the bagasse was milled and pretreated with diluted acid. The substrates were subjected to enzymolysis. The influence of particle size on reducing sugar conversion *via* pretreatment and enzymolysis was analyzed.

2.Experimental

2.1. Materials and Pretreatment

The bagasse was provided by a factory. The raw materials were washed, dried, milled, and screened. The chemical composition of the different bagasse particles was determined and listed in Table 1.

Table 1. Composition of differently sized bagasse particles

Particle Size (Mesh)	Glucan (%)	Xylan (%)	Klason Lignin (%)	Acid-Soluble Lignin (%)
10 - 20	49.44	25.39	18.23	0.30
20 - 40	50.99	25.04	20.03	0.29
40 - 60	50.15	24.69	22.85	0.28
60 - 80	50.05	23.65	23.14	0.30

2.2. Dilute Acid Hydrolysis

The dilute acid hydrolysis process was conducted in the electrical heated digester. The samples was treated at 160℃ for 15min with the solid liquid ratio of 1:15 and the H_2SO_4 concentration of 0.8%. After hydrolysis, the hydrolyzed samples were filtered through a 400-mesh cloth. The filtrate was tested for pH and stored at 4°C. The residuals were washed and dehydrated to 30% consistency and stored in a self-sealed plastic bag.

2.3.Enzymolysis

Enzymolysis reactions were conducted in a 50-ml Erlenmeyer flask with 1 g of the pretreated residuals, 100 ml of buffer (citric acid and sodium citrate buffer solution, 0.05mol/L, pH 4.8), and 20 FPU of cellulase enzyme. The flasks were stirred for 48 h at 50 °C and 200 rpm. After, the sample flasks were placed in a boiling water bath for 10 min to stop enzymolysis. Then, the samples were cooled in ice water and centrifuged for 5 min at 12000 rpm. The reducing sugar content of the supernatant was measured. The enzymolyzed substrates were washed, air-dried, and stored in plastic bags.

2.4.Reducing Sugar Analysis

The total reducing sugar content in hydrolyzed filtrates and enzymolyzed supernatants was calculated by ion chromatography. The conversion ratio of reducing sugar was calculated as follows:

$$C_{RSA}(\%)=W_1/(W_0 \times C_0) \times 100 \quad (1)$$

$$C_{RSA}(\%)=W_2 \times Y_1/(W_3 \times C_0) \times 100 \quad (2)$$

$$C_{TRS}\ (\%) = C_{RSA}\ (\%) + C_{RSE}\ (\%) \quad (3)$$

where C_{RSA} refers to the conversion ratio of reducing sugar in acid hydrolysis, C_{RSE} refers to the conversion ratio of reducing sugar in enzymolysis, C_{TRS} refers to the conversion ratio of total reducing sugar, C_0 refers to the total reducing sugar content in raw materials, W_0 refers to the weight of raw materials, W_1 refers to the reducing sugar weight in hydrolyzed filtrate, W_2 refers to the reducing sugar weight in enzymolyzed supernatant and W_3 refers to the weight of enzymolysis samples.

2.5.X-ray Diffraction (XRD) Analysis

Crystal structures in the different bagasse particles were determined by an X-ray diffractometer. The relative crystallinity (C_r) was calculated using the Segal formula (Eq. 1):

$$C_r\ (\%) = (I_{002} - I_{am})/I_{002} \times 100 \quad (4)$$

where I_{002} is the diffraction peak intensity of the 002 area and I_{am} refers to the scattering intensity of non-crystalline background with a diffraction angle of 18.

3.Results and Discussion

3.1.Effect of Bagasse Particle Size on Reducing Sugar Conversion

The sugar yield and the related conversion of RSA and RSE were listed in table 2 and table 3, separately.

Table 2. Sugar yield in hydrolyzed filtrates and enzymolysis supernatants

Particle	Hydrolyzed Filtrate					Enzymolysis Supernatant		
Size	Xylose	Glucose	Arabinose	Galactose	RSA	Xylose	Glucose	RSE
10-20	20.12	1.02	1.14	1.03	23.31	2.49	30.30	32.79
20-40	20.19	1.15	1.22	1.04	23.60	2.46	30.08	32.54
40-60	18.64	0.96	1.24	0.94	21.78	1.61	28.22	29.83
60-80	18.42	0.80	1.36	0.72	21.30	1.40	26.01	27.41

Table 3. Conversion ratio of RSA, RSE, and TRS

Particle Size	RSA	RSE	TRS
10-20	31.3	30.2	61.5
20-40	30.9	29.4	60.3
40-60	29.1	27.9	57.0
60-80	28.9	26.4	55.3

As showed in table 3, RSA and RSE conversion declined with decreased bagasse particle size. Since the smaller biomass particles allow better degradation, which releases more reducing sugars was dissolved.[11] However, the reducing degree was not very high. For example, the RSA conversion rate decreased from 31.3% to 28.9% when the bagasse particle size decreased from 10-20 mesh to 60-80 mesh. The smaller bagasse particles may have lower xylan content and more resistance from lignin (Table 1). Chen *et al.*[12] found that when the pretreatment time was less than 3 min, the reaction mainly happened on the milled bagasse surface. Thus, for small bagasse particles with high specific surface areas, the conversion ratio of reducing sugar was high.

In addition, when the bagasse particles were small, the corresponding conversion ratio of RSA and RSE was also small. The RSA and RSE conversion ratios of 10-20 mesh and 20-40 mesh particles were approximately 30 ± 0.4%. The RSE conversion ratio for 40-60 mesh and 60-80 mesh particles were 27.9% and 26.4% respectively, which decreased noticeable from the previous two samples. The maximum conversion ratio of was 61.5% in the 10-20 mesh samples, which mainly depended on RSE conversion.

3.2.Effect of Particle Size on the Lignin Content of Hydrolyzed Residuals

Figures 1 and 2 present the lignin content and its effects on reducing sugar conversion. As a filling and binding agent, lignin combines with and reinforces cellulose fibers *via* physical and chemical bonds in the cell wall.[13] As for the three components of lignocellulose, hemicellulose is the most easily degraded, while lignin is the most difficult to degrade.[14] But, nucleophilic molecules in lignin (such as the C_1 and C_6 of benzene ring) reacted with nucleophilic reagents and dissolved[15] during sulfuric acid treatment. The maximum RSE conversion ratio dropped 3.8% with increasing lignin content.

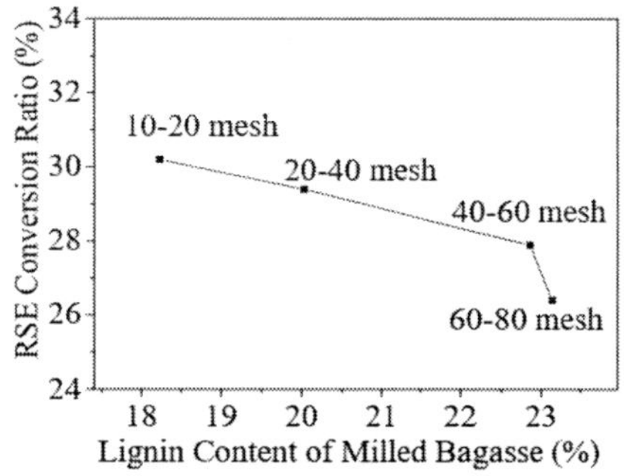

Figure 1. Effect of klason lignin content in milled bagasse on the RSE conversion ratio

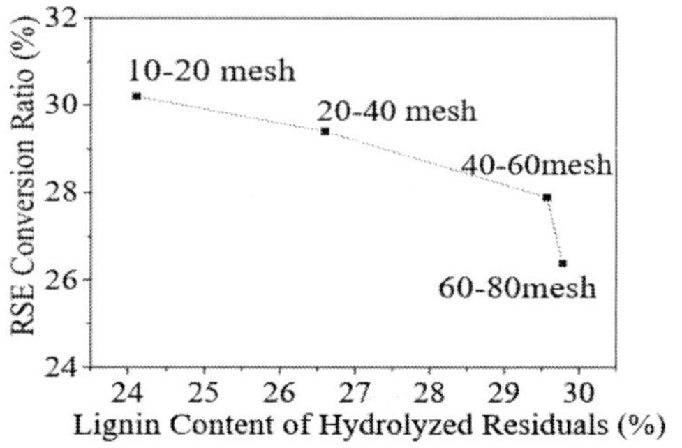

Figure 2. Effect of klason lignin content in hydrolyzed residuals on the RSE conversion ratio

Yu *et al.*[16] proposed that lignin prevented cellulase attachment to cellulose resulting in the low efficiency of enzymolysis. Masarin *et al.*[17] found that RSE conversion is inversely proportional to lignin content in hydrolyzed residuals. Lindedam *et al.*[18] also found that RSE conversion is proportional to carbohydrate content. In agreement with these results, when the lignin content of hydrolyzed residuals was high in this study, the RSE conversion ratio was low, probably because the lignin inhibited cellulase adsorption.[19]

3.3. Effect of Particle Size on Hydrolyzed Residuals Crystallinity

From the figure 3, it can be seen that the crystal structure of the four kinds of hydrolyzed residuals was similar. They were all composed by the crystalline region and amorphous area the same as milled bagasse particle. The crystallinity of milled bagasse particles was 37.0%, while the crystallinity of 10-20 mesh, 20-40 mesh, 40-60 mesh, and 60-80 mesh hydrolyzed residuals was 47.1%, 46.5%, 43.9%, and 43.6%, respectively. The increased relative crystallinity was due to the partial hydrolysis of amorphous cellulose and complete of hemicellulose hydrolysis.[20] The crystallinity of the 10-20 mesh and 20-40 mesh particles was similar, as was the 40-60 mesh 60-80 mesh particles, with the former set being about 3% higher than the latter. This result suggests that in a certain size ranges, milled bagasse particles react similarly to pretreatment.

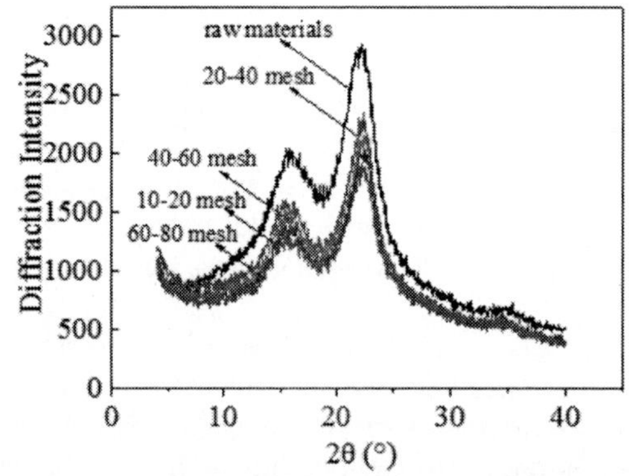

Figure 3. XRD spectra of milled bagasse particles after acid hydrolysis

The crystallinity of the 10-20 mesh hydrolyzed residuals was the highest, and the RSE conversion ratio was also the highest (30.2%). Similarly, the crystallinity and RSE conversion ratio (26.4%) of 60-80 mesh hydrolyzed residuals were the lowest. Higher crystallinity indicated that hemicellulose hydrolysis during pretreatment was better, which increased the accessibility of cellulose and improved the relative crystallinity of cellulose. It meant that when the crystallinity of the hydrolyzed residuals were higher, the enzymolysis was not more difficult[21]. Therefore, the efficiency of enzymatic hydrolysis was influenced by many factors including the crystallinity of hydrolyzed residuals.

Conclusions

1. With decreasing bagasse particle size, the conversion ratio of RSA and RSE declined. The conversion ratio of RSA fell from 31.3% to 28.9%. The conversion ratio of RSA and RSE was not increased by reducing the size of milled bagasse particles.
2. Cellulose and the lignin showed little degradation during hydrolysis. The size of milled bagasse particles was smaller, and the klason lignin in hydrolyzed residuals was higher, which decreased the reducing sugar conversion ratio. The optimal size of milled bagasse particles was 10-20 mesh, which produced a TRS conversion ratio of 61.5%, including 31.3% RSA conversion and 30.2% RSE conversion ratio.
3. Hydrolyzed residuals produced from 10-20 mesh particles showed the highest crystallinity and RSE conversion ratio. After acid hydrolysis, most hemicellulose was degraded and dissolved. The lignin was also degraded. These changes were very beneficial to enzymolysis.

Acknowledgements
The authors acknowledge the support of the Special Support Plan for High-Level Talent Cultivation of Guangdong Province (No.2014TQ01N603), Key Laboratory of Pulp and Paper Science & Technology of Ministry of Education of China (No.KF201508) and the National Water Pollution Control and Treatment Science and Technology Major Project (No. 2014ZX07213001).

References

[1] M. Irfan, M. Gulsher, S. Abbas, Q. Syed, M. Nadeem and S. Baig, *Sonklanakarin Journal of Sci. Technol.*, **33**, 397 (2001).
[2] D. Ciolacu, F. Ciolacu and V. I. Popa, *Cellulose Chem. Technol.*, **45**, 13 (2011).
[3] M. J. Taherzadeh and K. Karimi, *International journal of molecular Sci.*, **9**, 1621 (2008).
[4] M. Pedersen, K. S. Johansen and A. S. Meyer, *Biotechnol Biofuels*, **4**, 1986 (2001).
[5] P. Alvira, E. Tomás-Pejó, M. Ballesteros and M. Negro, *Bioresource Technol.*, **101**, 4851 (2010).
[6] A. Kumar, L. Singh and S. Ghosh, *Bioresource Technol.*, **100**, 3293 (2009).
[7] S.Y. Leu and J. Zhu, *Bioenergy Research*, **6**, 405 (2013).
[8] J. F. Kadla and R. D. Gilbert, *Cellulose Chem. Technol.*, **34**, 197 (2000).
[9] Y. Zheng, Z. Pan and R. Zhang, *International Journal of Agricultural and Biological Engineering*, **2**, 51 (2009).
[10] W. H. Chen, Y. J. Tu and H. K. Sheen, *Applied Energy*, **88**, 2726 (2011).
[11] J. Zhu and X. Pan, *Bioresource technology*, **101**, 4992 (2010).
[12] L. Chen, H. Zhang, J. Li, M. Lu, X. Guo and L. Han, *Bioresource Technol.*, **177**, 8 (2015).
[13] F. Kaya, J. Heitmann and T. Joyce, *Cellulose Chem. Technol.*, **33**, 203 (1999).
[14] H. Yang, R. Yan, H. Chen, C. Zheng, D. H. Lee and D. T. Liang, *Energy & Fuels*, 20, 388 (2006).
[15] W. Ji, Z. Shen and Y. Wen, *BioEnergy Research*, **7**, 1392 (2014).
[16] Q. Yu, X. Zhuang, Z. Yuan, W. Qi, W. Wang, Q. Wang and X. Tan, *Bioresource Technol.*, **144**, 210 (2013).
[17] F. Masarin, D. B. Gurpilhares, D. C. Baffa, M. H. Barbosa, W. Carvalho, A. Ferraz and A. M. Milagres, *Biotechnol Biofuels*, **4**, 55 (2011).
[18] J. Lindedam, S. B. Andersen, J. DeMartini, S. Bruun, H. Jørgensen, C. Felby, J. Magid, B. Yang and C. Wyman, *biomass and bioenergy*, **37**, 221 (2012).
[19] R. Huang, R. Su, W. Qi and Z. He, *Biotechnology progress*, **26**, 384 (2010).
[20] D. F. Caulfield and W. E. Moore, *Wood Sci For Prod Res Soc [Madison]*, 1974.
[21] S. Kim and M. T. Holtzapple, *Bioresource Technol.*, **97**, 583 (2006).

IOP Conf. Series: Materials Science and Engineering **205** (2017) 012007 doi:10.1088/1757-899X/205/1/012007

Biocomposites Based on Cellulose Acetate and 12-Aminolauric Acid Modified Montmorillonite

AS Ching[1] and LQ Reyes[1*]

[1]School of Chemical Engineering and Chemistry, Mapua Institute of Technology, Manila, Philippines
*lqreyes03@gmail.com

Abstract. Polymer-clay composites films based on cellulose acetate (CA) and 12-aminolauric acid-modified montmorillonite (AMMT) were prepared by solution intercalation method with clay loadings 1, 5, 10, 15 to 20%wt. Cellulose acetate is a good material resource due to its green qualities and ease with which it is processed, and using this material in combination with naturally derived material, montmorillonite, is a good means tune and improve the materials properties. To achieve compatabilization between the polymer and the clay, 12-aminolauric acid (12-ALA) was ionically exchanged with the Na^+-counter ions of the clay, producing an organically modified clay. The 12-ALA modifier is also an attractive resource due to its biocompatibility and relative abundance. The composites were characterized based on their chemical properties by Fourier Transform Infrared (FTIR), thermal properties by Differential Scanning Calorimetry (DSC), morphological characteristics by Scanning Electron Microscopy (SEM). X-ray diffraction (XRD) analyses revealed the presence of polymer intercalated and/or completely exfoliated silicate structures within the composites.

1. Introduction

The improvements in the functional properties of natural and semi-synthetic polymers have been made by reinforcement of the polymer matrix with layered silicates [1], [2]. Polymer-layered silicate (PLS) nanocomposites, where high aspect ratio silicate platelets of clay are dispersed in the polymer matrix. The formation of nanocomposites result to the improvement of the physical properties of these materials including increased mechanical strength, chemical resistance, gas barrier performance and flame retardancy when compared to the pure polymer or conventional composites [3]. Polymer nanocomposites with various clays have been extensively investigated due to their cost effectiveness and potential applications in different fields, including the structural, electronics, aerospace, and biomedical industries, and for use in sensors [4]. In the present study, the Na^+-Montmorillonite (Na^+-MMT) clay was used as a reinforcing material of the cellulose matrix. The attractiveness of MMT to be applied in many industries are due it its naturally abundant, nontoxic, chemically and thermally stable. It has been the most commonly employed layered silicate clay mineral in polymer composites and has been shown to improve the barrier and mechanical properties [5]. Also, cellulose acetate (CA) [6] is another material that is used in many applications because of their excellent processability. MMTs that are intercalated with fatty acid-derived surfactants was also used in the present study to promote the interxmixing between CA and MMT. Fatty acid based organo-modifier is an underutilized resource but offers many advantages in producing a polymer miscible nano-filler. The natural abundance, numerous, biocompatibility and thermal stability are some of the desirable characteristic of the fatty acid based modifiers [7]. The present study investigated the viability of preparing a CA based nanocomposites with different organo-clay loadings with different morphology, thermal stability, microstructures and tensile properties. To the knowledge of the author, this is the first study that utilized a fatty acid-modified clay as a filler of a cellulose acetate based nanocomposite.

2. Methodology

2.1. Intercalation of 12-ALA in Na^+-MMT

The 12-ALA modified MMT was synthesized based on Na^+-MMT (Nanofill 116, Southern Clay Products, Inc., Gonzales, Texas USA) and 12-ALA (Trade Tokyo Chemical Industry, Japan) by a simple cation exchange reaction. Briefly, twenty-five (25) grams of Na^+-MMT was dispersed to 1000 mL of deionized water at 90°C. The dispersion was stirred for approximately 1 hour. Independently, an appropriate amount of 12-ALA (2.0 x CEC of Na^+-MMT) was introduced in a 150 mL of pre-heated deionized water (90°C) with a pH of 1.5. The 12-ALA solution was slowly added to the Na^+-MMT slurry and was stirred for 2 hours at 90°C. The intercalated MMTs were harvested in the mixture by centrifugation at 2000 rpm, washed 10 times with deionized water and vacuum dried at 90°C for 24 hours.

2.2. Synthesis of Cellulose Acetate Composites

Pristine cellulose acetate (average $M_n \sim 30{,}000$ by GPC, Sigma-Aldrich) was purchased from Sigma-Aldrich and cellulose acetate composites consists of varying amounts of AMMT (0, 1, 5, 10, 15, and 20 wt.%) were prepared by means of solution intercalation method using tetrahydrofuran (THF) as solvent. The sample code assigned for the composite for example is CA-1AMMT which stands for the cellulose acetate with 1 wt % AMMT, and so on. Pristine CA serves as the control. Basically, 1.5 g of cellulose acetate along with the corresponding weight of AMMT were dissolved in 20 ml of THF under vigorous stirring for 1 hour at room temperature. Solvent casting method was used to prepare the polymer-clay composite films. The solution was placed on petri dish and air dried overnight.

2.3. Instrumentation

The characteristic functional groups of the samples were determined by Diffuse Reflectance Fourier-Transform Infrared Spectroscopic (DR-FTIR) analyses using a Shimadzu IR Prestige 21 Fourier-Transform infrared spectrophotometer (FTIR). The samples were scanned from 4000 to 600 cm^{-1}. Phase transformations and melting transitions were determined by Differential Scanning Calorimetry (DSC) using a TA Instruments DSC Q10. The analysis was performed at a heating rate of 10°C/min in N_2 gas with temperature range from 30 to 300°C. The melting temperature (T_m) was acquired in the thermograms. Morphological changes of the clay material, pristine and organo-modified clay were observed by using a Hitachi S-3400N Scanning Electron Microscope at an accelerating voltage of 10kV. To determine the intercalation, molecular structure, and conformation of 12-ALA in the clay in its silicate layers, and the formation of intercalated polymer-clay composite films, samples were subjected to x-ray diffraction analyses with a Shimadzu XRD-700 x-ray diffractomer (XRD) with CuKa radiation ($\lambda = 1.5406$ Å) at 40kV and 30 mA in the 2θ range of 3-90° with a scan rate of 2°/min. Values of interplanar (d) spacing were calculated according to Bragg's equation.

3. Results and Discussion

Polymer-clay composites of cellulose acetate and 12-aminolauric acid-modified clay was developed by solution intercalation method. Initially, the organo-modified clay was synthesized via cation exchange between the Na^+ ion of MMT and 12-ALA. The Na^+ ions of the clay material was determined to be exchanged with 100% × CEC surfactant by thermogravimetric analysis. The modification of the clay material with cationic surfactant result to the significant changes in their physical and chemical properties such as: hydrophobicity, basal spacing, textural properties and surface morphology and even biocompatibility.

3.1. Fourier-Transform Infrared (FTIR) Spectroscopic Analyses

The functional groups and possible molecular interaction between CA and AMMT were evaluated by FTIR. The FTIR spectra of pristine CA and samples of CA with varying amounts of AMMT are shown in Figure 3.1. For CA, the band at 2944 cm^{-1} represents the aliphatic of C-H group. The band at 1751cm^{-1} is attributed to the stretching vibration of C=O bond and a band at 1370 cm^{-1} shows O-H group. The characteristic bands around 1433.56 cm^{-1} represent the bending vibration of CH_2, 1238 cm^{-1} assigned stretching vibration of the C-O group and the band at 902 cm^{-1} corresponding to the out-of-plane C-H bend. In addition, the main characteristic bands of CA are slightly shifted, which may be due to weak polymer-clay interaction or overlapping characteristic bands of CA and AMMT. It should be noted that no peak is observed to indicate the presence of a chemical bond between AMMT and CA chains. CA-AMMT composites are considered as effective biosorbents due to their high content of amino acid and hydroxyl functional groups, which show significant adsorption potential for the removal of various aquatic pollutants.

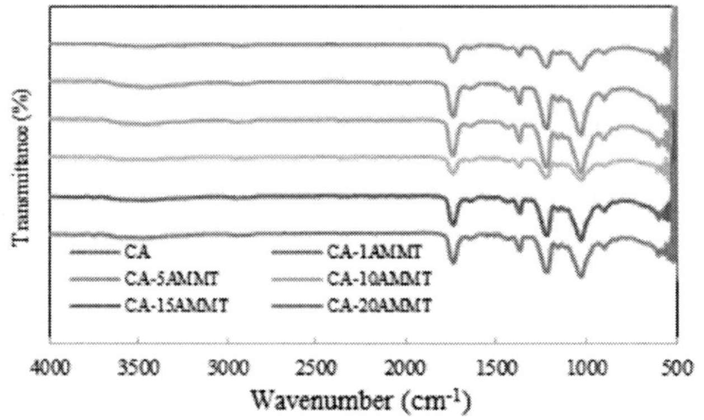

Figure.1. FTIR spectra of CA and its grafted forms

3.2. Thermal Analyses of Cellulose Acetate/ Organo-clay Composites

The melting behaviors of free 12-ALA and the clay-bound surfactant were probed by DSC and is shown in Figure 3.2. The 12-ALA surfactant in free form exhibited a sharp melting peak at 189°C which indicates that the melting confined alkylamine. Additionally, the T_m of the surfactant in AMMT was lower than that of the pristine 12-ALA as shown in Figure 3.2 and 3.3. The order of the alkyl amine in the silicate sheets is affected by the combination of the electrostatic attraction and repulsion between alkyl chains silicate sheets and carboxylic acid groups. It has been pointed out earlier that the amino fatty acid may be ionically bound to the silicate sheets of the clay or attracted to the metal ion via ion-dipole interaction.

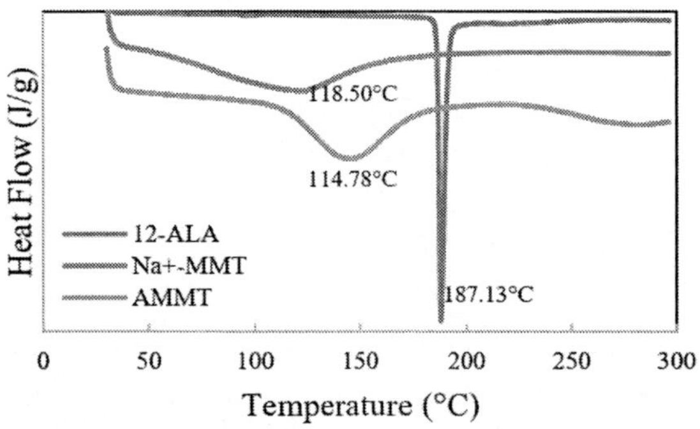

Figure.2. DSC thermogram of 12-ALA, Na^+-MMT and AMMT.

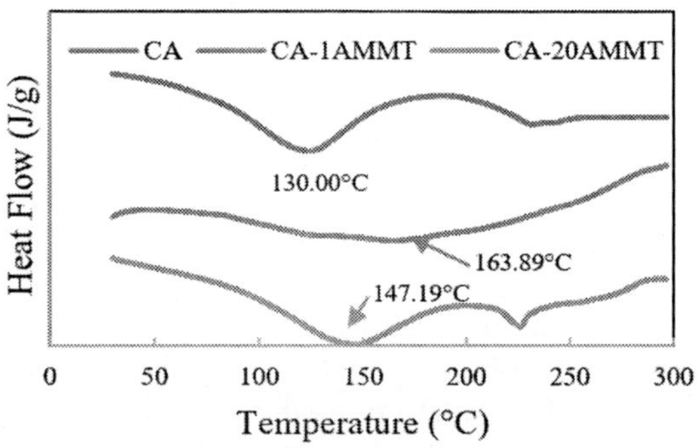

Figure.3. DSC thermogram of CA, CA-1AMMT AND CA-20AMMT.

The thermal transition of CA and the polymer-clay composites were probed by DSC as shown in Figure 3.3. The thermogram of CA exhibited a sharp endothermic transition at 130°C which is attributed to the phase transition of the polymer from solid to liquid. The polymer melting temperatures in the composite were found to vary with clay loading. It can be observed that CA with AMMT composites exhibited higher T_m. This might be attributed to the overall reduction in crystallite size with increasing organo-modified clay loading. The size of these microstructures also affects polymer melting temperature. The broadening of the melting endotherm might suggest the reduction of the polymer crystallites where the clay silicate layers acts as physical barriers during the crystallization process.

3.3. Morphological Analysis of Polymer-Clay Composites

SEM was employed to examine the morphological properties of 12-ALA, unmodified Na^{+}-MMT, 12-ALA-intercalated MMT (200CEC-AMMT), pure CA, and the CA-AMMT composite films. Their electron micrographs are depicted in Figure 3.4. It can be observed that the unmodified MMT was smooth and non-agglomerated (Figure 3.4b).

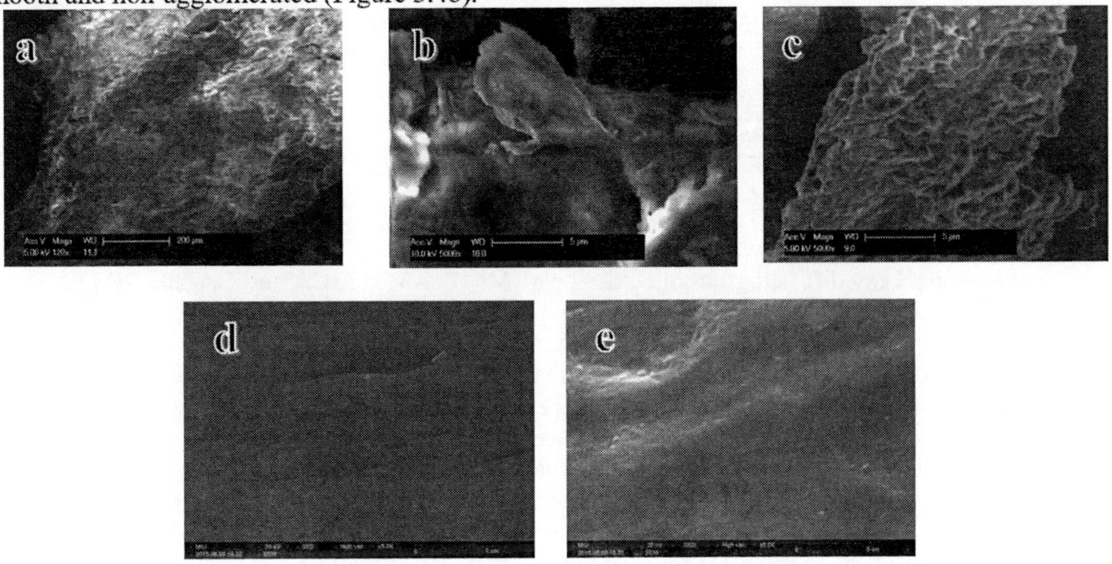

Figure.4. Scanning Electron Micrographs of (a) 12-ALA, (b) Na^{+}-MMT, (c) AMMT, (d) CA-AMMT and (e) CA-20AMMT at 3,500X.

Interestingly, a significant change in geometry and morphology for the organo-modified clays were observed. The AMMT clay (Figure 3.4c) exhibited massive and highly aggregated flakes that exhibit rough surface and edges. The unmodified MMT, initially, exhibits curved plates which then transform into flat flakes depending on the surfactant loading. The transformation of the clays plates is heavily affected by the molecular interaction with the surfactant head group and tail group to the silicate sheet and the interaction of the surfactant tail group with another surfactant tail group. The micrographs of CA-20AMMT composite films (Figure 3.4d and e) presented a more rugged structure on the surface, in which agglomeration clay indicates good adhesion at the polymer-clay interface.

3.4. Microstructural Analysis of Polymer-Clay Composites

X-ray diffraction analysis revealed the presence of intercalated surfactants within the clay's alumino-silicate galleries. The XRD patterns of Na^{+}-MMT and its organo-modified forms are shown in Figure 3.6. It was observed that with 12-ALA loading in the clay, there is a shift of the (001) peak to a lower 2θ position indicating that there is an increase in d_{001}-spacing. The 2θ position was found to be at 4.85° when the 12-ALA was loaded to the unmodified clay. The AMMT used in the study is a pseudotrilayered structure, the calculated d_{001}-spacings should be 1.75 nm. However, the d_{001}-spacing observed varies with that of calculated, and so it can be deduced that the 12-ALA does not accurately/precisely adopt the mentioned conformation. These d_{001}-spacing deviations can be accounted to the fact that the molecular dynamic calculation assumed that the methylene (CH_2) in the hydrocarbon chain of 12-ALA adopts an all-anti conformation, discounting the presence of methylenes in gauche/disordered conformation and coexistence of 12-ALA isomers in same clay system. Initially, Na^{+}-MMT exhibited a narrow and sharp diffraction peak of the 001 position; then, at

12-ALA loading, the 001 diffraction peaks became sharper and more intense which suggests the formation of more ordered intercalated surfactant. The pristine CA and CA-20AMMT composite films were also subjected to X-ray diffraction. The diffraction peak of AMMT might be masked by the diffraction peaks of the polymer. The diffraction planes attribute to cellulose acetate were (101), (002), and (10̄1), found at 21, 25, 29°, respectively. The 001 diffraction plane of AMMT was not observed for the CA-20AMMT which may indicate that the former is exfoliated in the polymer matrix.

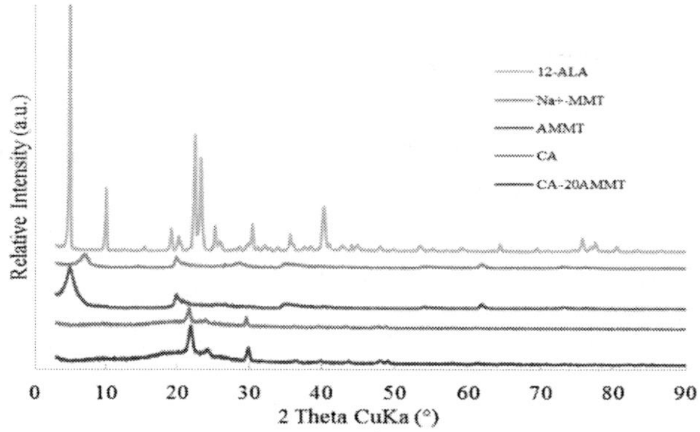

Figure5. X-ray diffractograms 12-ALA, Na^{+}-MMT, AMMT, CA and CA-20AMMT.

4. Conclusion

Polymer-clay composite (CA-AMMT) films, with varying loads of 12-ALA-modified montmorillonite (AMMT) as the organo-modified clay, were successfully prepared by solution casting method. The organo-modified clay was successfully synthesized by a simple cation exchange in which the Na^{+} counter ions of the clay are exchanged with the protonated 12-ALA. The formation of composites was assessed by X-ray Diffraction. XRD analyses showed a broadening/disappearance of diffraction peak of the AMMT, indicating the presence of intercalated and exfoliated layers of AMMT. Due to the controlled swelling of the material, the polymer-clay composites may be used as food packaging, gas barrier and wound dressing material. Another potential use of the polymer-clay composites is in controlled drug delivery and tissue engineering applications.

5. References

[1] S. Adeosun, G. Lawal, S. Balogun, and E. Akpan, "Review of Green Polymer Nanocomposites," *Scirp.Org*, vol. 11, no. 4, pp. 483–514, 2012.

[2] C. Williams and M. Hillmyer, "Polymers from Renewable Resources: A Perspective for a Special Issue of Polymer Reviews," *Polym. Rev.*, vol. 48, no. 1, pp. 1–10, 2008.

[3] M. Hetzer and D. De Kee, "Wood/polymer/nanoclay composites, environmentally friendly sustainable technology: A review," *Chem. Eng. Res. Des.*, vol. 86, no. May, pp. 1083–1093, 2008.

[4] G. Fan, M. Wang, C. Liao, T. Fang, J. Li, and R. Zhou, "Isolation of cellulose from rice straw and its conversion into cellulose acetate catalyzed by phosphotungstic acid," *Carbohydr. Polym.*, vol. 94, no. 1, pp. 71–76, 2013.

[5] N. Kaur and D. Kishore, "Montmorillonite: An efficient, heterogeneous and green catalyst for organic synthesis," *J. Chem. Pharm. Res.*, vol. 4, no. 2, pp. 991–1015, 2012.

[6] S. Tungprapa, T. Puangparn, M. Weerasombut, I. Jangchud, P. Fakum, S. Semongkhol, C. Meechaisue, and P. Supaphol, "Electrospun cellulose acetate fibers: Effect of solvent system on morphology and fiber diameter," *Cellulose*, vol. 14, no. 6, pp. 563–575, 2007.

[7] P. Čapková, M. Pospíšil, M. Valášková, D. Měřínská, M. Trchová, Z. Sedláková, Z. Weiss, and J. Šimoník, "Structure of montmorillonite cointercalated with stearic acid and octadecylamine: Modeling, diffraction, IR spectroscopy," *J. Colloid Interface Sci.*, vol. 300, no. 1, pp. 264–269, 2006.

IOP Conf. Series: Materials Science and Engineering **205** (2017) 012008 doi:10.1088/1757-899X/205/1/012008

Silver-Loaded Cellulose Acetate-g-Poly(ε-caprolactone) Composites

CR Tuburan[1], LE Dela Rosa[1] and LQ Reyes[1]

[1]School of Chemical Engineering and Chemistry, Mapua Institute of Technology, Manila, Philippines
Email : lqreyes03@gmail.com

Abstract. Cellulose acetate (CA) was grafted with poly(ε-caprolactone) PCL oligomers via the ring-opening of ε-caprolactone (ε-CL) monomer initiated by the hydroxyl functionality of CA. The incorporation of short PCL oligomers in CA's structure caused the transformation of it crystalline domains into amorphous phases (internal plasticization) as observed by differential scanning calorimetry (DSC). Another evidence of plasticization induced by grafting was the significant reduction of the degradation temperature and stiffness of the copolymers. Proton Nuclear Magnetic Resonance (^{1}H-NMR), Fourier-Transform Infrared (FTIR) Spectroscopies and Gel Permeation Chromatography (GPC) verified success the grafting as suggested by the attachment of PCL on the glucose ring and increase in polymer molecular weights after the reaction. Due to the good films forming ability of the synthesized CA grafted with PCL (CA-g-PCL) material, it was loaded with silver nitrate ($AgNO_3$) and the composite was observed to be have bactericidal against a gram negative bacteria, *Escherichia coli*, and a gram positive bacteria, *Bacillus subtilis*.

1.Introduction

Biodegradable polymers propose an appealing remedy to the environmental problem of plastic waste accumulation and disposal; however, their properties such as poor mechanical properties, high permeability to gases and poor processibility have limited their use to a narrow variety of applications. [1]. The natural polysaccharides cellulose and starch are excellent candidate for making of low-cost materials for high technology applications and to improve their physical, chemical and processing properties of these materials, low molecular weight additives is required (external plasticization) [2], [3]. On the other hand, covalently linking lactones and carboxylic acids to these polysaccharide chain is one method of improving their processing properties (internal plasticization). This technique introduces defects in the crystalline domains and disrupts inter and intrachain hydrogen bonding within the polysaccharide structure [2]. One of the most commonly explored polysaccharide derivative for a myriad of applications is cellulose acetate (CA), synthesized by the reaction of cellulose with acetic anhydride [4]. While the acylation process improves the processibility of cellulose (ie. solubility in many organic solvents), the resulting material has high stiffness and brittleness, agin limiting its application [3]. And so, the rigidity of CA can be tuned by grafting aliphatic polyesters to the hydroxyl group of CA by chemical means. The biodegradable polyester poly(ε-caprolactone) (PCL) has been used as a plasticizer to CA producing a material with characteristics better than their individual polymers. Nevertheless, the high price of PCL has prevented its widespread industrial use. The development of polymer graft is an alternative to reduce costs and tune material properties [3].

Polymers have been modified with other types of materials for them to have multifunctional properties, including the development of antimicrobial polymer composites. There is an effort, nowadays, for the use of inorganic nanoparticles as a potent antimicrobial compounds to relieve the issue of drug resistance. Among the various inorganic metal nanoparticles, silver nanoparticles have received considerable interest for two reasons: (1) silver is a successful antimicrobial agent; and (2) it demonstrates excellent biocompatibility human cells [5]. To the knowledge of the researchers, this is the first study that exhaustively monitored the occurrence of grafting of PCL to CA and the effect of grafting to the thermal and mechanical properties of the CA-g-PCL. The present study also demonstrated a facile method of grafting PCL in CA where the chemical reaction is achieved in 15 minutes. Cellulose acetate has also received attention for biotechnological applications due to its ease in forming films, fibers, textiles and nanoparticles. Similar to CA, the developed CA-g-PCL materials can easily be casted into films. The researchers took the opportunity to embed silver, the form of silver nitrate, to develop an antimicrobial composite film that were tested against *Staphylococcus aureus* and *Escherichia coli.* .

2.Methodology

2.1. CA-graft-PCL synthesis

PCL-graft-CA polymers with weight/volume ratios of 12.5%, 25%, 37.5%, and 50% CA:ε-CL were carried out in polymerization set up equipped with a magnetic stirring bar. The polymer products were designated with sample codes based in the ration between CA and PCL. For example, the polymer product with the code CA_{50}-g-PCL, means that the CA charged in the reactor is 2.5grams with 5mL of ε-caprolactone (ε-CL) in the present of 0.1 wt% catalyst, stannous (II) octoate. Then, this mixture was reacted at 180°C using a thermostatic oil bath for 15 mins under reduced pressure and continuous nitrogen gas purging. Then composites were then further purified by dissolving the polymer products with a small amount of toluene and then precipitating them in hot methanol to remove the unreacted CL and short chain PCL; that may have possibly formed during the reaction. The extracted sample was dried in a vacuum at 60°C for 24 hours. 10 wt/vol% polymer solutions were prepared using THF as solvent. The solutions were casted into petri dishes and the solvent was allowed to evaporate for 24 hours. Polymer/silver composite was prepared by mixing 10 wt% of $AgNO_3$ to 10 % wt/vol of graft copolymers (in THF). The CA-graft-PCL/Ag solution was stirred for 2 hours. The composite solution was then casted on a Petri dishes, and the solvent was allowed to vaporize for 24 hours. Then, the copolymer film was vacuum dried for another 4 hours at 80°C. The film was cut into circular discs using a paper puncher. The discs were UV-sterilized for 1 hour before antimicrobial evaluation. The sample was in a desiccator until use.

2.2. Characterization of Cellulose Graft Copolymers

The molecular weight of the samples was determined by a Simadzu HPLC chromatograph apparatus equipped with a Shimadzu refractive index (RI) detector. The samples were eluted with tetrahydrofuran (THF) through a TSK-GEL G3000HHR at a flow rate of 1 mL/min. The molecular weights were determined based on polystyrene standard and the calibration was made using polystyrene standards. NMR spectra of the products were obtained using a JEOL Lambda FT-NMR spectrometer. Samples (10 to 20 mg) were dissolved in deuterated chloroform (CDCl3). The functional groups present in the polymer product were determined using an IR Prestige 21 Fourier Transform Infrared Spectrophotometer. Diamond attenuated total reflection was used for the samples in powder form. The sample was scanned 10 times from 4000 to 600 cm^{-1}. DSC thermograms were acquired using a differential scanning calorimeter (TA Q10 series). The analysis was performed by heating the samples at 10°C/min in N_2 The mechanical properties of the composites were evaluated using an Instron Single Column Universal Testing Machine (UTM). The samples were cut into 40mm x 10mm strips and their Young's modulus and %tensile strength at break were measured in triplicate. CA-g-PCL & CA-g-PCL/Ag composite films were tested for their antimicrobial activity by the Kirby–Bauer disk diffusion method against *Bacillus subtilis* and *Escherichia coli.* The bacteria were cultured in Mueller Hinton agar plates for 24 hours. Then, a bacterial colony was introduced to

Mueller-Hinton (MH) broth and was allowed to multiply. The disc samples were placed on each quadrant of the plate that was streaked, previously, with the bacteria. Culture plates with bacteria were then incubated at 35°C for 24 hours.

3.Results and Discussion

3.1. Synthesis of CA grafted with PCL

Cellulose acetate grafted with poly(ε-caprolactone) oligomers was successfully synthesized by ring opening polymerization of ε-CL initiated by the free hydroxyl terminal of the CA. The of the grafted CA was prepared by solvent free ring-opening polymerization of CL. Residual hydroxyl groups of CA was used to initiate the polymerization as shown in Scheme 3.1

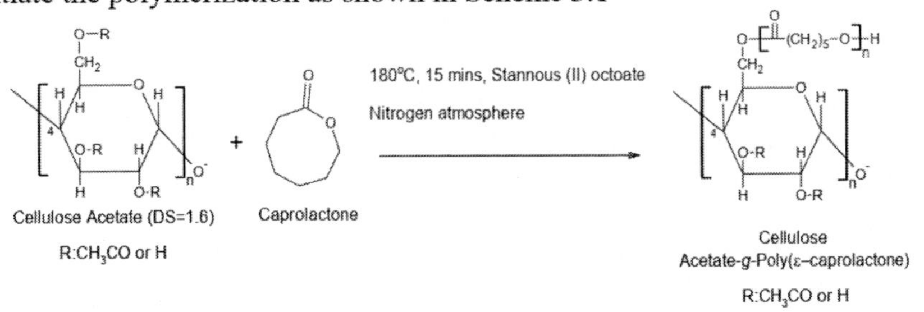

Scheme 3.1. Plasticization of cellulose acetate via ring opening polymerization of PCL.

3.2. Molecular weight determination after grafting

Table 3.1. Molecular weight pristine CA and CA-g-PCLs.

Sample Code	M_n	M_w	PDI
CA	45,520	109,900	2.41431
CA_{50}-g-PCL	51,093	113,547	2.22233
$CA_{37.5}$-g-PCL	85,291	139,637	1.63719
CA_{25}-g-PCL	61,163	121,554	1.98739
$CA_{12.5}$-g-PCL	97,631	148,520	1.52124

GPC was employed to measure the resulting increase in the polymers' number average (M_n) and weight average molecular weights (M_w). After grafting, the polymer products exhibited increase in M_n and M_w that suggests that PCL chain are grafted on the CA backbone. Molecular weights of the pristine and modified CA are listed in Table 3.1. The molecular weights of CA-g-PCL were found to be in the range 110,000 to 150,000 g/mol. In addition, it is observed that the molecular weight of the polymer grafts increased with the decreasing CA content. It should be considered that CA contains the hydroxyl initiation sites for ROP and lower CA during the reaction allows the PCL chains to elongated only on few actives sites. Also, fewer initiations sites of polymerization promoted monodispersity (lower PDI) of the polymer molecular weight.

3.3. Analysis Molecular Structure of the ε-CL grafted CA

The chemical structures of CA and the copolymer is quite similar; however, the occurrence of grafting could be verified by FTIR analysis. The IR spectra of the grafted and pristine CA samples can be depicted in figure 3.1. It can be seen that the carbonyl group's vibration can be intensively seen at 1735 cm^{-1}. It is important to see the vibration of the increasing $-CH_3$ moiety of the acetyl groupw in the chain. The presence of this aliphatic carbon chains could indicate the success of the ring opening polymerization reaction from the CL, which could be found at 2943 and 2870 cm-1. The decrease in the OH vibration at 3468.01 cm^{-1} between the neat and modified cellulose acetate is almost unnoticeable. The reason for this could be that the vibration from the free hydroxyl

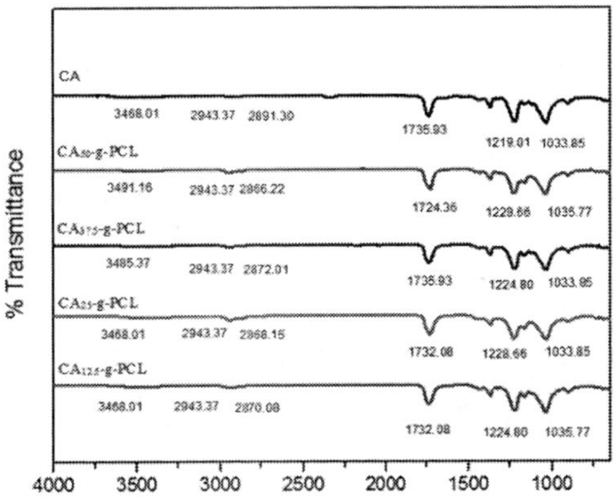

Figure 3.1. IR Spectra of pristine CA and its grafted forms.

groups of unmodified cellulose acetate would not change on esterification because each PCL chain contains also one –OH group at its end.

The structure of the grafted polymer was analyzed in detail by NMR spectroscopy. The ^{1}H-NMR spectra of the CA grafted with PCL is shown in Figure 3.2. The methyl groups of the substituent acetyl groups appear at the region of 2.2–1.7 ppm, and pyran ring protons resonances arising from the seven ring protons of H_1–H_6 in each anhydrous glucose unit appear in the region of 5.5–3.2 ppm. Furthermore, the signals from the ring protons overlap with each other, and the accurate assignment of these resonances has never been performed. In addition, the characteristic peaks of the PCL chain were also observed in the spectra with the chemical shifts at 1.30, 1.54-1.60 and 2.30 ppm which are attributed γ-, overlapping β- and δ-, and α-methylenes, respectively. Again, the peaks from the PCL units are clearly present in the range of δ 1.2–4 ppm and imply that there is an occurrence of ring-opening polymerization reaction of CL. Resonances from the cellulose peaks are evident as small, broad peaks in the region δ 3.4–3.8 and 5.0–5.4 ppm. The results agree well with a previous report [3] that the signal of the hydrogen attached to the aliphatic $-CH_2-$ group appear at 1.7 ppm, those related to the ester group at 2.3 and 4.0 ppm, and the hydrogens of the $-CH_2-$ group neighboring the chain end –OH group are detected at 3.4 ppm.

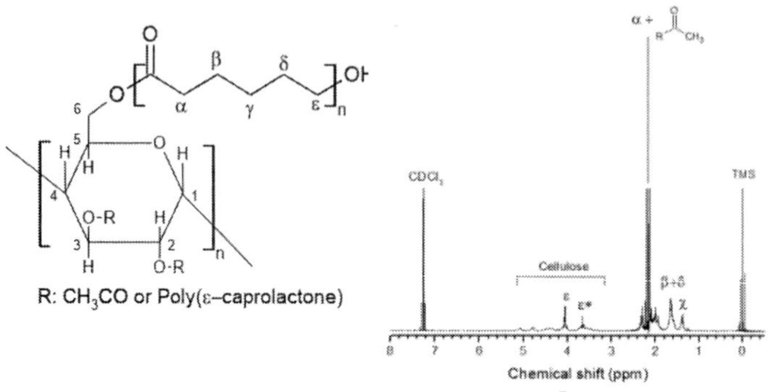

Figure 3.2. (a) Molecular structure of CA-*g*-PCL with its corresponding hydrogen positions, and (b) ^{1}H-NMR spectra of representative modified cellulose acetate.

3.4. *Thermal properties of the PCL grafted CA*

The melting temperature (T_m) of the grafted polymer composite ranged from 45°C to 50°C with enthalpy of melting (ΔH_m) at 2-3J/g. Interestingly, CA is a semi-crystalline polymer but, with the presence of PCL, CA becomes an amorphous polymer. This may be attributed to the added intermolecular interaction and repeating unit stereo irregularity of the monomer.

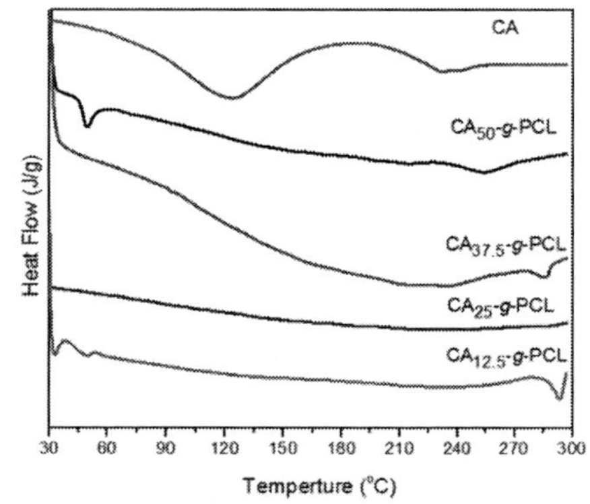

Figure 3.3. DSC thermograms of pristine CA and grafted CA

3.5. *Tensile properties of the CA-based materials*

Plasticization is assumed to increase the flexibility of chains and lead to a decrease in polymer stiffness. The load-extension curves of the modified CA are presented in Figure 3.4. Pristine CA exhibited the highest modulus of elasticity (876 MPa), derived from the load-extension curve. Significant decrease in moduli observed for the CA-g-PCL products. For example, $CA_{12.5}$-g-PCL exhibited a reduction in Young's modulus to 492 MPa, indicating that pristine CA exhibits high stiffness while CA-g-PCL products is less stiff. This is might be due to the introduction of imperfection on the lattice of CA by the bulkiness of PCL chain, creating void spaces and leading to the decrease in the modulus.

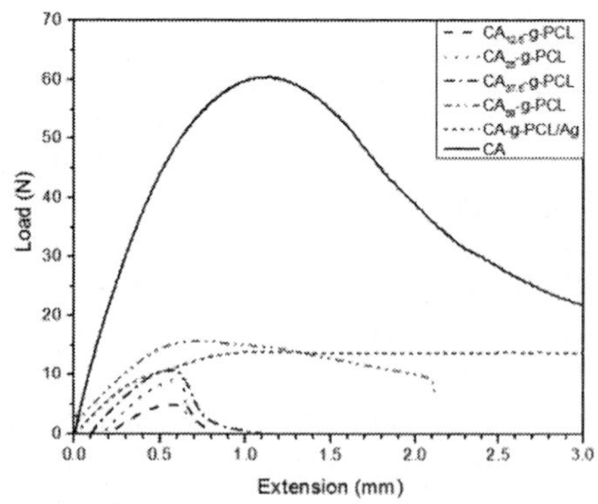

Figure 3.4. Load-extension curves of pristine and grafted CA.

IOP Conf. Series: Materials Science and Engineering **205** (2017) 012008 doi:10.1088/1757-899X/205/1/012008

3.6. Evaluation of Bactericidal Property of CA-g-PCL loaded with Ag

Silver particles is a potent antimicrobial material and when incorporated in a polymeric matrix may offer its property to the latter. The Kirby-Bauer disc diffusion assay was conducted to observe the broad antimicrobial activity of silver that is loaded on CA-based polymers. The antibacterial activity of the materials was tested against *Escherichia coli*, a gram negative bacterium and *Bacillus subtilis*, a Gram-positive, rod-shaped bacterium. Figure 3.6 presents the growth of the microorganism on the surface of the agar. Unloaded polymer films Figure 3.6, b, d and e did not inhibit the growth of bacteria where they are placed however, silver-loaded CA-g-PCL exhibited little zone of no growth within its periphery for both microorganisms even if the silver loaded is 10% of the polymer's weight.

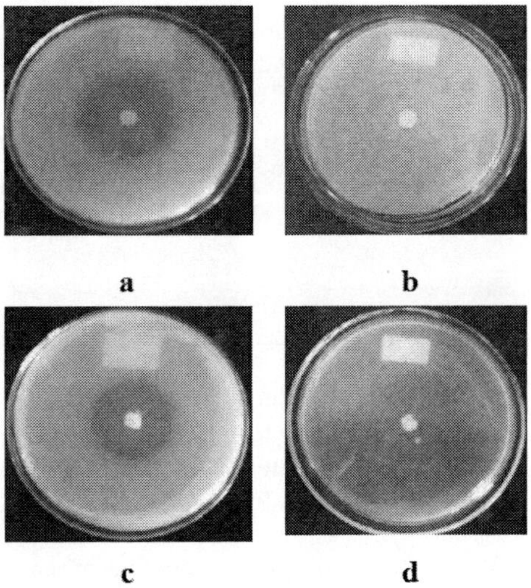

Figure 3.5. Photographs of bacterial cultures after incubation with the composite materials. (a) silver-loaded $CA_{37.5}$-g-PCL and (b) negative control against *E. coli.* (c) silver-loaded $CA_{37.5}$-g-PCL and (f) negative control against *B. subtilis*.

4.Conclusion

The present study demonstrated a facile way of synthesizing a graft copolymer of cellulose acetate (CA) and poly(ε-caprolactone) (PCL). The ring opening of the e-CL monomer is believed to be initiated by the hydroxyl of CA chain and their subsequent polymerization was initiated by the hydroxyl terminal of the growing PCL chain. The grafting of PCL chains to cellulose acetate resulted in internal plasticization. The potential application of the CA-g-PCL as a matrix for antimicrobials was observed by embedding silver in the graft copolymer and antimicrobial efficacy against both gram-negative and gram-positive bacteria. The overall grafting procedure may find many applications as antimicrobial packaging, wound dressing textile, or drug carrier.

5.References

[1] B. D. Ulery, L. S. Nair, and C. T. Laurencin, "Biomedical applications of biodegradable polymers," *J. Polym. Sci. Part B Polym. Phys.*, vol. 49, no. 12, pp. 832–864, 2011.

[2] C. Yan, J. Zhang, Y. Lv, J. Yu, J. Wu, J. Zhang, and J. He, "Thermoplastic cellulose-graft-poly(L-lactide) copolymers homogeneously synthesized in an ionic liquid with 4-dimethylaminopyridine catalyst," *Biomacromolecules*, vol. 10, no. 8, pp. 2013–2018, 2009.

[3] G. Számel, A. Domján, S. Klébert, and B. Pukánszky, "Molecular structure and properties of cellulose acetate chemically modified with caprolactone," *Eur. Polym. J.*, vol. 44, no. 2, pp. 357–365, 2008.

[4] H. N. Cheng, M. K. Dowd, G. W. Selling, and A. Biswas, "Synthesis of cellulose acetate from cotton byproducts q," *Carbohydr. Polym.*, vol. 80, no. 2, pp. 449–452, 2010.

[5] A. J. Huh and Y. J. Kwon, "'Nanoantibiotics': A new paradigm for treating infectious diseases using nanomaterials in the antibiotics resistant era," *J. Control. Release*, vol. 156, no. 2, pp. 128–145, 2011.

IOP Conf. Series: Materials Science and Engineering **205** (2017) 012009 doi:10.1088/1757-899X/205/1/012009

Adsorption of Ni(II) onto Chemically Modified Spent Grated Coconut (*Cocos Nucifera*)

F I Hamzah[1], **K Khalid**[2] **and M A K M Hanafiah**[2]

[1] Faculty of Bioresource and Food Industry, Universiti Sultan Zainal Abidin 22000 Besut, Terengganu, Malaysia,
[2] Faculty of Applied Sciences, Universiti Teknologi MARA, Cawangan Pahang, 26400 Bandar Tun Razak Jengka, Pahang, Malaysia

Abstract. A new adsorbent of plant waste origin from coconut processing food factory was explored for removing Ni(II) from aqueous solutions. Several parameters such as pH, dosage, concentration and contact time were studied to obtain optimum conditions for treatment of Ni(II) contaminated wastewater. Spent grated coconut (*Cocos nucifera*) treated with sulfuric acid (SSGC) showed good adsorption capacity for Ni(II) ion. The amount adsorbed was affected by solution pH with the highest value achieved at pH 5. Other optimum conditions found were; dosage of 0.02 g, and 60 min of equilibrium time. Ni(II) adsorption obeyed the pseudo-second order kinetic model which suggested that chemisorption mechanism occurred in the adsorption process. The equilibrium data presented a better fitting to the Langmuir isotherm model, an indication that monolayer adsorption occurred onto a homogeneous surface. The maximum adsorption capacity, q_{max} was 97.09 mg g^{-1}, thus SSGC can be classified as good and comparable with other plant waste adsorbents.

1. Introduction

Coconut tree (*Cocus nucifera*) is one of the members of Arecaceae (palm family) that grows especially in the tropics and is considered as one of the important crops in Malaysia. It is easily grown and can be found in different types of soil such as loamy, laterite, coastal sandy and reclaimed soils. Coconut is a versatile plant, having a multipurpose source of food and non-food products. In Malaysia, coconut waste is easily obtained due to its major consumption by local community. Spent grated coconut (SGC) from a defatted coconut residue is a by-product of a coconut milk factory. After the extraction of coconut milk, SGC fiber by-product finds other applications such as fertilizers and animal feed [1].

The term heavy metal is referring to any metallic element presence in periodic table which has a relatively high density, toxic and mutagenic effect at low concentration. The excessive release of heavy metal into the environment due to industrialization activities has become a major source of water pollution. The chemical exposure happens when sources such as factory, incinerator, tank, plant, landfill and chemical store released toxic level of chemicals to the surrounding and come in contact with human. Zinc, lead, cadmium, mercury, chromium and nickel are the known type of heavy metals which are commonly found in wastewater discharge. Due to their non-biodegradable and persistent in nature, heavy metals can accumulate in the soil and thus harm the ecosystem. Heavy metals enter human bodies via three pathways; ingestion, inhalation, and absorption through our large organ (skin) and mucous membrane and will be stored in human body tissues [2]. If exposed to human, heavy metals will be accumulated in body tissues, bound to proteins, impair enzymatic activity, slowly poison organ and ultimately result in fatality [3]. Ni(II) is one of the heavy metals of concern and the maximum permissible in aqueous solution is 0.02 mg L^{-1} based on WHO guidelines [4].

In wastewater removal technique, adsorption is employed to remove heavy metal ions at low concentrations which is better than ion-exchange, electrolysis, chemical precipitation, membrane separation, carbon adsorption and co-precipitation which are used at higher concentrations. In this experiment, spent grated coconut was used as an adsorbent because it is quite cheap, low economic value and easy to obtain from local market. Most of the adsorption studies have been focused on plant waste such as corn silk [5], rape straw [6], sugarcane bagasse [7], lalang (*Imperata cylindrical*) leaf powder [8], sugarcane bagasse pith [4] and husk of *Lathyrus sativus* [9].

The aim of this work was to evaluate the adsorption potential of sulfuric acid treated coconut spent grated for Ni(II) removal from aqueous solutions. Experiments were done in a batch system to study the effect of initial pH, biosorbent dosage and initial Ni(II) ion concentration. The adsorption was modelled by kinetic models to understand the mechanism while the isotherm models were used to model the adsorption equilibrium.

2. Methodology

2.1. Chemical treatment of spent grated coconut (Cocos nucifera)

The sample of spent grated coconut was collected from a food factory in Kuching, Malaysia. The pre-treatment and modification methods were adapted from our recent study [1]. The spent grated coconut was dried in an oven at 105 °C overnight and designated as SGC. It was washed with hexane (R&M Chemicals, UK) in the ratio of 1.0 g SGC to 10 mL of hexane (1:10, w/v) and was stirred using magnetic stirrer for 16 h. This method removed fat and eliminated the organic interferences in sample matrix. The pre-treated sample was filtered and dried at 70 °C for 3 h. The hexane-washed SGC was labelled as HSGC. Acid treatment was then carried out to enhance the adsorption. Sulfuric acid (H_2SO_4) in particular is widely used where it is able to improve pore development. The dehydrating effect of a concentrated H_2SO_4 is known to reduce the carbon atoms in a plant. The modification involved mixing 30 g of HSGC with 500 mL (3.0 M) H_2SO_4 (ratio of 3:50, w/v) at 100 °C. The mixture was stirred continuously at 480 rpm for 3 h, cooled to room temperature before washed twice with 1 L deionized water each time. The slurry was then filtered and soaked in 1 L (1.5% w/v) $NaHCO_3$ solution for 24 h to remove excess acid. Next, the sample was washed thoroughly with distilled water until the pH of the effluent was close to 7. The samples were dried for overnight in an oven at 105 °C, ground and sieved to obtain average particle size of 120-250 μm. The H_2SO_4 modified powder was labelled as SSGC

2.2. Nickel batch adsorption experiment

All batch adsorption experiments were carried out in duplicates and the results were reported as average. All chemicals used were of analytical reagent grade. The standard solution of Ni(II) (1000 mg L^{-1}) was purchased from Fluka (Germany). Experimental solutions of the desired Ni(II) concentrations were obtained by successive dilutions from the standard Ni(II) solution. For adsorption experiments, a known weight of the adsorbent was mixed with 50 mL Ni(II) solutions in a conical flask at room temperature (30.0±0.5)°C, shaken at 120 stroke/min for a certain period of time. The initial pH of Ni(II) solution was fixed at 5 by adding 1 M hydrochloric acid (HCl) or sodium hydroxide (NaOH) solutions.

The effect of pH was studied over a pH range of 2-5 to avoid precipitation of $Ni(OH)_2$. To study the effect of dosage, the weight of adsorbent was varied from 0.01 to 0.10 g. The effects of Ni(II) concentration, contact time and kinetics were evaluated by shaking SSGC with Ni(II) solutions (5, 10 and 30 mg L^{-1}) at various time intervals (5-120 min). The isotherm study was conducted by equilibrating Ni(II) solutions (5-80 mg L^{-1}) with 0.02 g SSGC. After adsorption, the mixture was filtered and the filtrate was analysed for Ni(II) content using an Atomic Absorption Spectrometer (PinAAcle 900T, PerkinElmer, USA). The amount of Ni(II) adsorbed, q_e, (mg g^{-1}) and percentage of removal were calculated by using equation (1) and equation (2) respectively:

$$q_e = \frac{C_o - C_e}{m} V \tag{1}$$

$$Removal(\%) = \frac{C_o - C_e}{C_o} x\,100\% \tag{2}$$

where, C_o (mg L^{-1}) and C_e (mg L^{-1}) are Ni(II) concentrations before and after adsorption, respectively; V (L) is the volume of Ni(II) solution, and m (g) is the weight of SSGC.

3. Results and discussion

3.1. Effect of pH

It is well recognized that the pH is an important parameter governing the adsorption of metal ions by sorbent and impacted the adsorption process. Figure 1 shows the change in pH contributed to the removal efficiency. Theoretically, the solution pH significantly affects the metal chemistry and the surface properties of adsorbent. The pH also affects the ionization states of functional groups, such as carboxyl and hydroxyl on cellulose-rich adsorbents.

The optimum pH was 5 with highest amount of Ni(II) adsorbed (q_e) of 23.43 mg g^{-1}. Less adsorption took place at the initial pH 2 (19.82%) that explained on the basis of active sites protonation, resulting in H^+ and Ni^{2+} competition to occupy the active sites. A continuous increase in q_e occurred at the pH range of 2-5. The increase in the q_e was due to reduction of the electrostatic repulsion between the surface of SSGC and Ni^{2+} ion in bulk liquid.

3.2. Effect of adsorbent dosage

From Figure 2, q_e decreased from 38.08 to 4.73 mg g^{-1} as dosage was increased from 0.01 to 0.1 g. The percentage removal of Ni(II) however increased from 76.16 to 94.67%. The q_e at low dosage was high because Ni(II) could easily access the adsorption sites. The q_e however decreases at higher amount of dosage which could be due to the aggregation of adsorbent particles resulting in a decrease in the effective sites for adsorption. Thus the optimum dosage was selected as 0.02 g of which it achieved the optimum percentage of removal and the amount of Ni(II) absorbed.

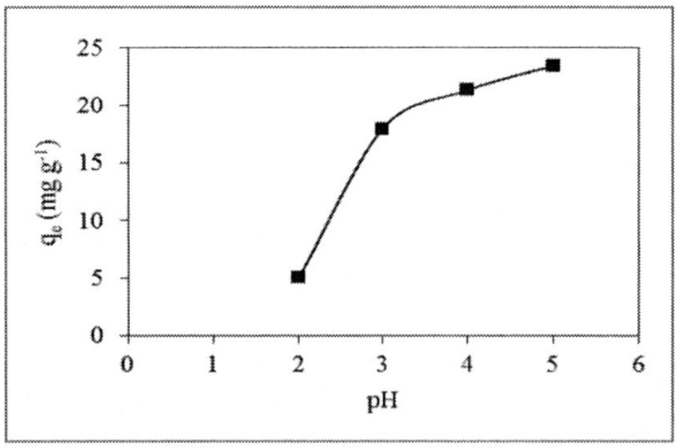

Figure 1. Effect of pH on Ni(II) adsorption onto SSGC (C_o: 10 mg L^{-1}; dosage: 0.02 mg).

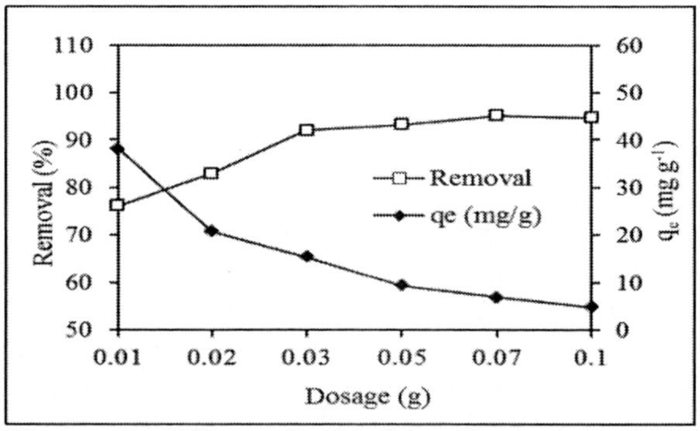

Figure 2. Effect of adsorbent dosage on Ni(II) adsorption onto SSGC (C_o: 10 mg L^{-1}; pH: 5).

3.3. Effects of nickel concentration, contact time and adsorption kinetics

Figure 3 shows the adsorption data at different initial Ni(II) concentrations and contact time. It can be clearly seen that the q_e and equilibrium time are dependent on the concentration on Ni(II). The q_e increased with contact time for all Ni(II) concentrations. The percentage removal of Ni(II) ions was initially increased due to the high availability of surface area of SSGC. The amount adsorbed remain unchanged and reached equilibrium within 30 min for Ni(II) concentrations of both 5 mg L^{-1} and 10 mg L^{-1}, and extended up to 60 min for Ni(II) concentrations of 30 mg L^{-1}.

The pseudo-first order and pseudo-second order models were used to examine the kinetics adsorption process. The pseudo-first order indicates the intraparticle diffusion model which adsorption is controlled by diffusion into the boundary layer [10] while pseudo-second indicates the chemisorption process. The correlation coefficient (R^2) can be evaluated and the adsorption capacities between calculated and experimental values can be compared. The expression of pseudo-first order [11] and pseudo-second order models [12] are given by equations (3) and (4) respectively:

$$\log (q_e - q_t) = \log q_e - \frac{k_1}{2.303} t \qquad (3)$$

$$\frac{t}{q_t} = \frac{1}{h} + \frac{1}{q_e} t \qquad (4)$$

The symbol of q_e indicates the amount of Ni(II) adsorbed (mg g^{-1}) at equilibrium, while q_t indicates that amount of Ni(II) adsorbed at time t (min), k_1 is the rate constant of adsorption process (min^{-1}) of the pseudo-first order, k_2 (g/mg.min) is the pseudo-second order rate constant while h (mg/g.min) is the initial adsorption rate and can be calculated by using the formula; $h=k_2q_e^2$.

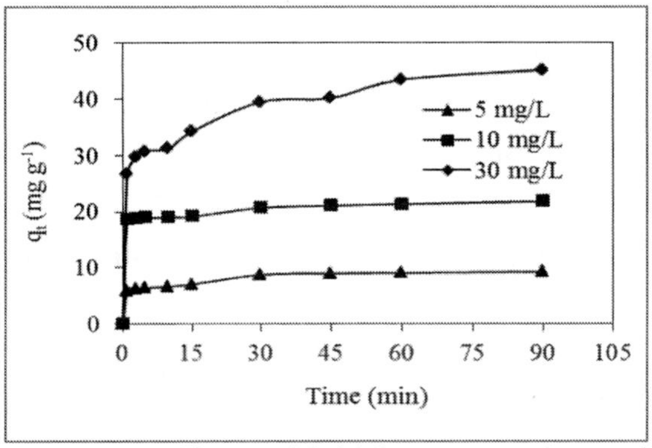

Figure 3. Effects of initial concentration and contact time on Ni(II) adsorption onto SSGC (pH 5; dosage: 0.02 mg)..

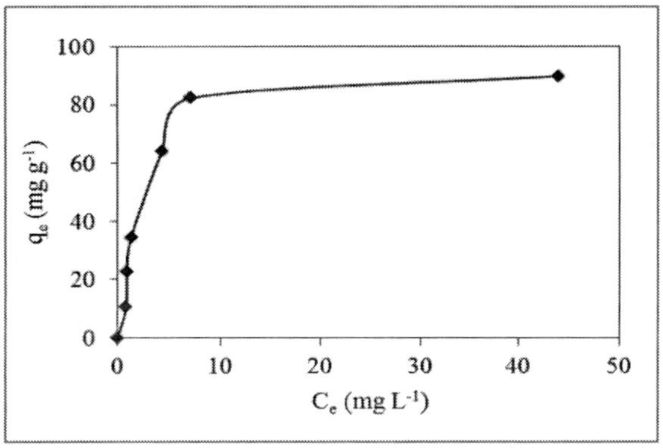

Figure 4. Isotherm plot of Ni(II) adsorption by SSGC (pH 5; dosage: 0.02 mg). .

Table 1 shows the pseudo-first order and pseudo-second order parameters at different Ni(II) concentrations. From this table, the R^2 values were between 0.867–0.969 while the rate constants (k_1) were in the range of 0.010 to 0.017 min^{-1} using pseudo-first order model. The low R^2 values for all concentrations suggested that it is less appropriate to use pseudo-first order expression. However, the R^2 values were much better for the pseudo-second order which were in the range of 0.954 to 0.999. From the calculations, the values of $q_{e, cal}$ were close to the values of $q_{e, exp}$. Thus the Ni(II) adsorption follows the pseudo-second order kinetic model which suggested the chemisorption process.

Table 1. Pseudo-first order and pseudo-second order parameters at different concentrations of Ni(II).

		Pseudo-first order			Pseudo-second order			
Ni(II) (mg L^{-1})	$q_{e, exp}$ (mg g^{-1})	$q_{e, cal}$ (mg g^{-1})	k_1 (min^{-1})	R^2	$q_{e,cal}$ (mg g^{-1})	k_2 (g/mg.min)	h (mg/g.min)	R^2
5	9.230	2.735	0.016	0.867	9.372	0.050	4.400	0.998
10	21.820	2.805	0.017	0.969	21.787	0.049	23.148	0.999
30	45.167	1.858	0.010	0.944	45.249	0.008	16.949	0.954

IOP Conf. Series: Materials Science and Engineering **205** (2017) 012009 doi:10.1088/1757-899X/205/1/012009

3.4. Adsorption Isotherm

The isotherm explains the relationship between the amount of Ni(II) adsorbed (q_e, mg g^{-1}) and the concentration of adsorbate in bulk solution (C_e, mg L^{-1}) at a given temperature under equilibrium conditions. The term adsorption equilibrium is identified when the rate of adsorbate adsorbed is equal to the rate of adsorbate being desorbed from the adsorbent. The adsorption isotherm results are shown in Figure 4. The values of adsorption capacities increased as the concentration of Ni(II) increased. The highest q_e observed for the Ni(II) adsorption onto SSGC was 89.91 mg g^{-1}.

The adsorption isotherm data was further analysed by using the Langmuir and Freudlich isotherm models. The Langmuir isotherm model indicates that there is a limited area available for adsorption, the adsorbate is adsorbed as a monolayer molecule. The Freundlich model meanwhile, assumes that the adsorbent surface sites have different binding energies which indicate the multilayer adsorption behaviour of adsorbate molecules onto the heterogeneous surface of the adsorbent [13]. The linearized Langmuir and Freundlich isotherm expressions are represented in equations (5) and (6) respectively:

$$\frac{C_e}{q_e} = \frac{1}{q_{max}\,b} + \frac{C_e}{q_{max}} \tag{5}$$

$$\log q_e = \log K_F + \frac{1}{n}\log C_e \tag{6}$$

where q_e is the amount adsorbed at equilibrium, q_{max} is the maximum adsorption capacity (mg g^{-1}), b is a constant (L mg^{-1}) related to the energy of adsorption which reflects the affinity between the adsorbent and the adsorbate, C_e is the equilibrium Ni(II) ion concentration (mg L^{-1}). K_F and n are the maximum adsorption capacity (mg g^{-1}) and adsorption intensity of the adsorbate ions on the adsorbent, respectively.

Table 2 presents the calculated parameters obtained from the linearized plots (not shown) of Langmuir and Freundlich models. The results revealed that the experimental equilibrium data had been in good agreement with Langmuir model. Based on R^2 value and the close values of $q_{e,\,exp}$ and $q_{e,\,cal}$., Langmuir model was favourable and suited the adsorption of Ni(II) onto SGCC. The high R^2 values indicated that the model was suitable for describing the tested system under the concentration range studied, suggesting that Ni(II) adsorption was limited to monolayer coverage. This indicates that the surface is relatively homogeneous in terms of functional group where interaction occured. However, the Freundlich coeffiecient of correlation value obtained for the Ni(II) adsorption on SSGC was lower (0.728) although the n value obtained was 2.17. The n value over 1 indicated the favourable adsorption [9]. These conclude that the adsorption data fitted the Langmuir model with q_{max} recorded as 97.09 mg g^{-1} better than the Freundlich model.

Table 2. Langmuir and Freundlich isotherm constants and correlation coefficients.

$q_{e,exp}$ (mg g^{-1})	Langmuir			Freundlich		
	q_{max} (mg g^{-1})	b (L mg^{-1})	R^2	K_F (mg g^{-1})	n	R^2
89.91	97.09	0.32	0.988	28.59	2.17	0.728

Adsorption capacities of other adsorbents for the removal of Ni(II) are given in Table 3 for comparison and showed that SSGC is an effective adsorbent for the removal of Ni(II) from aqueous solutions.

Table 3. Comparison of adsorption capacities by various adsorbents.

Adsorbent	Solution pH	C_o (mg L^{-1})	q_{max} (mg g^{-1})	Reference
Sugarcane bagasse (untreated)	5	10-200	2.23	[7]
NaOH treated *Imperata cylindrica*	5	5-50	6.96	[8]
Husk of *Lathyrus sativus*	5	10-1000	15.70	[9]
Sugarcane bagasse pith (untreated)	6.5	50-1000	73.56	[4]
SSGC	5	5-80	97.09	This study
Sugarcane bagasse pith (activated carbon)	6.5	50-1000	140.85	[4]

4. Conclusions

This study revealed that the optimum pH was reached at pH 5.0 and the adsorbent dosage of 0.02 g. The initial adsorption rate was fast and reached equilibrium in less than 60 min. The adsorption kinetics followed the pseudo-second order kinetic model which suggested the chemisorption process. It has been found that one gram of the SSGC could adsorb 97.09 mg Ni(II) and the adsorption process was best fitted to Langmuir isotherm. Thus, this work showed that SSGC is a potential adsorbent for the removal of Ni(II) from aqueous solutions.

Acknowledgment

This work was supported by UiTM Research Grant Scheme (Dana: 600-RMI/ST/DANA 5/5/Dst (179/2011)

References

[1] Khalid K and Hanafiah M A K M 2014 *Adv Mater Res.* **970** 192–7.
[2] Song Q and Li J. 2015 *Environ. Pollut.* **196** 450–61.
[3] Demirbas A 2008 *J. Hazard. Mater.* **157(2–3)** 220–9.
[4] Anoop K K, Sreejalekshmi K G and Baiju R S 2011 *Bioresour Technol.* **102(22)** 10239–47.
[5] Yu H, Pang J, Ai T and Liu L 2016 *J. Taiwan Inst. Chem. Eng.* **62** 21–30.
[6] Wu Y and Wang L 2016 *Procedia Environ. Sci* **31** 75–80.
[7] Alomá I, Martín-Lara M A, Rodríguez I L, Blázquez G and Calero M. 2012. *J. Taiwan Inst.Chem. Eng.* **43(2)** 275–81.
[8] Hanafiah M A K M, Zakaria H and Wan N W S 2010 *Clean - Soil, Air, Water* **38(3)** 248–56.
[9] Panda G C, Das S K, Bandopadhyay T S and Guha A K 2007 *Colloids Surf B Biointerfaces* **57(2)** 135–42.
[10] Piccin J S, Gomes C S, Feris L A and Gutterres M. 2012 *Chem Eng J.* **183** 30–8.
[11] Ho Y S and Mckay G 1998 Process Saf. Environ. Prot. **76** 332–40.
[12] Ho Y S, Ng J C Y and McKay G 2000 Sep. Purif. Rev. **29(2)** 189–232.
[13] Srivastava S, Agrawal S B and Mondal M K 2015 *Ecol. Eng.* **85** 56–66.

IOP Conf. Series: Materials Science and Engineering **205** (2017) 012010 doi:10.1088/1757-899X/205/1/012010

Recovery Of Chromium Metal (VI) Using Supported Liquid Membrane (SLM) Method, A study of Influence of NaCl and pH in Receiving Phase on Transport

Muhammad Cholid Djunaidi*, Retno A. Lusiana, Maya D. Rahayu

Chemistry Departement of Diponegoro University, Indonesia
Email: cholidjunaidi@live.undip.ac.id

Abstract. Chromium metal(VI) is a valuable metal but in contrary has high toxicity, so the separation and recovery from waste are very important. One method that can be used for the separation and recovery of chromium (VI) is a Supported Liquid Membrane (SLM). SLM system contains of three main components: a supporting membrane, organic solvents and carrier compounds. The supported Membrane used in this research is Polytetrafluoroethylene (PTFE), organic solvent is kerosene, and the carrier compound used is aliquat 336. The supported liquid membrane is placed between two phases, namely, feed phase as the source of analyte (Cr(VI)) and the receiving phase as the result of separation. Feed phase is the electroplating waste which contains of chromium metal with pH variation about 4, 6 and 9. Whereas the receiving phase are the solution of HCl, NaOH, HCl-NaCl and NaOH-NaCl with pH variation about 1, 3, 5 and 7. The efficiency separation is determined by measurement of chromium in the feed and the receiving phase using AAS (Atomic Absorption Spectrophotometry). The experiment results show that transport of Chrom (VI) by Supported Liquid membrane (SLM) is influenced by pH solution in feed phase and receiving phase as well as NaCl in receiving phase. The highest chromium metal is transported from feed phase about 97,78%, whereas in receiving phase shows about 58,09%. The highest chromium metal transport happens on pH 6 in feed phase, pH 7 in receiving phase with the mixture of NaOH and NaCl using carrier compound aliquat 336.

1. Introduction

Chemical and metallurgical industries are the main producers of metal-bearing liquid effluents. Due to the toxic character of several metals contained in the metal-bearing wastewater, it needs to be treated before final discharge to the environment. Thus, industries are increasingly being forced either by the administration, regulatory, cost pressures and social pressures to reduce the amount and environmental sensitivity of the liquid waste they produce [1].

Chromium has fairly high toxicity value and potentially as harmful pollutant when waste industries is not processed first and directly discharged into the environment. The concentration of chromium (III) and (VI) in electroplating liquid waste or metal coating industry is generally higher than the threshold values allowed. The maximum amount allowable of total chromium present in healthy water is 0.05 mg L^{-1}. However, chromium metal is also useful for people in daily activities as well as in industrial processes, such as for electroplating [2].

Toxic but valuable of this metals in another hand, causes the separation and recovery of waste containing chromium are essential. Recycling and re-use is an important aspect in the development of the industry in the future.

One of the methods for separating metal ions is liquid separation membrane [3]. In the liquid membrane technique, carrier compounds play an important function. Carrier compound as a facilitator is determinant in the separation performance from the feed phase. A carrier compound should has high ability of extraction via stable complex formation within the membrane, high selective separation towards certain species, good solubility as well as diffusion coefficient in the suitable organic solvent (membrane) and can be used in relatively small amounts [4]. In addition, the selectivity of carrier compound towards certain metal ions is determined by the active groups present on the carrier compound.

There are several methods that can be used for the separation and recovery of metals using Liquid Membrane including Supported Liquid Membrane (SLM), Emulsion Liquid Membrane or ELM and Bulk Liquid Membrane or BLM. BLM is used for separation of Cr(III), Cu(II) and Cd(II) with carrier compound that has active atom N and S so that the transport efficiency was Cd(II)>Cu(II)>Cr(III) based on HSAB (Hard Soft Acid Base) theory [5].

SLM is a separation technique consisting of two homogen phases that are feed phase and receiving phase separated by semipermeable membrane. The advantages of SLM are a little amount of carrier compound required, simple and inexpensive operation [6]. SLM system contains three main components: a supporting membrane, organic solvents and carrier compounds. The separation with SLM is largely determined by the selectivity and effectivity of the compounds (carrier). Type of carrier compound determines how much ionic metal can be transported to the receiving phase.

Solangi, et al (2013) [7] in his research reported that SLM is efficient in transport Cr(VI) using calixarene carrier and NaOH receiving solution. Whereas Khasanah (2001) [8] used SLM to subtract the Cl anion from sea water desalination with the supported membrane PTFE impregnated with the carrier compound aliquat 336 with the desalination percentage of about 96,69% for aliquat compound 336 and 100% for aliquat compound 336-TBP on pH 6 in receiving phase.

Based on the information above, this research uses method of SLM with the supported PTFE membrane and aliquat carrier compound 336 in karosen. The pH variation on feed phase are about 4, 6, 9 and 1, 3, 5, 7 on receiving phase. The solution used on receiving phase are HCl, NaOH, HCl-NaCl and NaOH-NaCl. Those variation aim to make Cr(VI) ion in electroplating liquid waste can be separated and to produce purer metal and separated from its matrics.

2. Experiment Setup

2.1 . Studying the Effect of pH receiving phase

To study the effect of pH in the receiving phase, we make various pH, namely, 1, 3, 5 and 7. Whereas feed phase uses the electroplating liquid waste with pH 4. The carrier compound used is aliquat 336 1M.

2.2. Studying the Effect of pH Feed phase

To study the effect of pH in the feed phase, the variation of pH is made i.e. 4, 6, and 9, while the receiving phase uses solution consisting of NaOH-NaCl 1M at pH 7. The carrier compounds used is Aliquat 336 1M.

2.3. Studying the Effect of NaCl consentration in receiving phase.

To study the effect of NaCl in the receiving phase, the variation of NaCl concentration are made, i.e. 0M, 0.5M, 1M and 1.5M in pH 7. Meanwhile, feed phase uses the electroplating waste with pH 6. The carrier compounds used is aliquat 336 1M.

2.4. Studying the Effect of pH in receiving phase with NaCl addition

To study the effect of NaCl addition in receiving phase with the variation of pH, the solution in receiving phase is made from HCl-NaCl mixture 1 M for pH 1, 3, 5; and NaOH-NaCl mixture 1 M for

IOP Conf. Series: Materials Science and Engineering **205** (2017) 012010 doi:10.1088/1757-899X/205/1/012010

pH 7. In the meantime, the feed phase uses the electroplating waste with pH 4 and the carrier compound used is aliquat 336.

3. Result and Discussion

The initial step is measurement of the levels of Cr (VI) in wastewater with AAS. The result shows that the wastewater contains 1055 mg/L of Cr (VI).

3.1. The effect of pH of receiving phase

To study the effect of pH of the receiving phase, a series of pH variation is made from HCl and NaOH mixture in different level: 1, 3, 5, and 7; meanwhile, the pH of feed phase is constant coming from the electroplating waste at pH 4. Furthermore, the aliquat 336 is used as a carrier compound in membrane phase. The result can be seen in table 1.

Table 1. The effect of pH of Receiving phase on Cr(VI) transport using aliquat 336 as carrier compound

pH receiving phase	% transport From Feed to Membrane Phase	% transport In Receiving Phase (From Membrane Phase)
1	61.86	6.61
3	57.65.	1.51
5	67.63	0.99
7	71.62	1.46

As seen in table 1, when the pH of receiving phase increases the transport of Cr(VI) from feed phase is also high; however, unfortunately low concentration can be transported into the receiving phase since Cr(VI) is trapped in the membrane phase. However, transport of Cr(VI) into receiving phase is low with the increasing pH of receiving phase and slightly becoming high when the pH is acid. In order to make alkaline (bases) condition in the receiving phase we use NaOH while use HCl is for making acid condition. This condition resulting in low transport of Cr(VI) into the receiving phase when the pH is high because OH- ion has smaller affinity than Cl-, consequently, Cl ions are more easily exchange with $HCrO_4^-$ ions which will be released into the receiving phase.

3.2. The influence of pH of Feed phase

In feed phase, the pH is varied by 4, 6 and 9 meanwhile in the receiving phase pH is constant at pH 7 by adding NaCl 1M as seen in table 2.

In the transport process, $HCrO_4^-$ ion that has already been bound in the membrane phase will be released into the receiving phase, while the anions in the receiving phase will be bound by the membrane. The addition of Cl^- in the receiving phase aims to replace $HCrO_4^-$ so that $HCrO_4^-$ will be released easily into receiving phase. The high concentration of $HCrO_4^-$ ions is found in receiving phase because the Cl^- has higher affinity to the aliquat 336 than to the OH^- ion [12].

Table 2. The relationship between % transport of Cr(VI) with variation of pH of feed phase in receiving phase (pH constant at 7 and NaCl 1M).

pH feed phase	% transport From Feed to Membrane Phase	% transport In Receiving Phase (From Membrane Phase)
4	50.36	22.59
6	97.78	46.56
9	91.29	40.35

3.3. The effect of NaCl concentration addition in receiving phase

To determine the effect of NaCl concentration added to receiving phase, the NaCl concentration is varied from 0M, 0.5M, 1M to 1.5M. The Effect of NaCl on transport in the receiving phase can be seen in table 3. It can be seen in table 3 that the higher the NaCl concentration, the higher Cr(VI) can be transported. Nevertheless, at NaCl 1.5M transport of Cr(VI) decreases due to the high osmotic pressure so that the carrier compound aliquat 336 is relased easily from membrane phase. The highest transport of Cr(VI) occurs at concentration of NaCl 1M with the percentage of Cr(VI) transport of about 97.78% in feed phase and 58.09% in receiving phase.

Table 3. The effect of NaCl concentration in receiving phase on transport of Cr(VI)

Concentration of NaCl (xM)	% transport From Feed to Membrane Phase	% transport In Receiving Phase (From Membrane Phase)
0	37.91	2.48
0.5	83.79	46.56
1	97.78	58.09
1.5	85.52	55.88

3.4. The Effect of pH of receiving phase with NaCl addition.

To compare the effect of pH and Cl^- ion on Cr(VI) transport, further study is conducted. The impact of pH variation of receiving phase with NaOH-NaCl 1M liquid can be seen at table 4. Table 4 shows that the addition of NaCl 1M in receiving phase affects the chromium metal transport. In comparison to the table 1 by using the same pH variation, chromium metal transport is high if NaCl solution is used in receiving phase. The highest transport of metal still occurs at pH 7 transporting metal from feed phase. On the one hand, it can be seen from table 1 that the higest transport of metal to receiving phase happens at pH 1. It may occur since at pH 7 there is no Cl^- ion resulting in the chromium metal is hard to release in receiving phase. Whereas, at pH 1 as shown it table 4 the concentration of NaCl is more than 1M with the result that the high osmotic pressure causes the aliquat 336 carrier compound aparts from the membrane, so that it leads to low metal transport.

Table 4. The effect of pH variation of receiving phase towards Cr(VI) transport with aliquat 336 as carrier compound and NaOH-NaCl as liquid receiving.

pH receiving phase	% transport from feed Phase to Membrane phase	% transport in Receiving phase
1	87.23	7.38
3	94.35	24.61
5	97.51	20.84
7	97.78	58.09

According to data from table 3 and 4, it can be concluded that metal chromium that can be transported using aliquat 336 as carrier compound in SLM is high because the concentration of anion (Cl^-) in the receiving phase increases chromium metal transport significantly.

4. Conclusion

This study shows that transport of Chrom (VI) by Supported Liquid membrane (SLM) is influenced by pH solution in feed phase and receiving phase. Furthermore, the NaCl addition in receiving phase increases the number of Cr (VI) in receiving phase.

1. The higest percentage of chromium metal is 97.78 % in feed phase and 58.09 % in receiving phase with aliquat 336 carrier compound.
2. The greatest chromium metal transport occur at pH 6 in feed phase and pH 7 on receiving phase with the mixture of NaOH and NaCl 1M.

References

[1] Agreda, D., de, I. Garcia-Diaz, F.A. López and F.J. Alguacil, 2011, *Revista de Metalurgia, 47 (2)*, 146-168.
[2] Choi, Y.M, Seung, H. Y., 2004, *Sep. Sci. Technol.* , **39**, 1663-1680

IOP Conf. Series: Materials Science and Engineering **205** (2017) 012010 doi:10.1088/1757-899X/205/1/012010

[3] Cleij,M.C., Scrimin, P., Tecilla, P., and Tonellato, U. , 1997, *J. Org. Chem.*, **62** (16), 5592-5599•
[4] Bartsch, R.A., dan Way, J.D., 1996, *Chemical Separation With Liquid Membrane, CS Symposium Series, 642.*, American Chemical Society, Washington, page 1-6
[5] Djunaidi, M.C., Khabibi, Ulumuddin, 2017. IOP Conf. Series. Materials and Engineering **172**. 012032
[6] Misra, B. M., 1996 *Chemical Separation With Liquid Membrane, CS Symposium Series, 642.*, American Chemical Society, Washington, 361-375
[7] Solangi, I. B., Ozcan, F., Arslan, G., Ersoz, M., 2013., *Sep. Purif. Technol..*,**18**, 470-478
[8] Khasanah , R., *Desalination of Sea Water Using Supported Liquid Membrane (SLM), Thesis*, Diponegoro University, 2001.
[9] De, Anil, K, 1997, *Solvent Extraction of Metal*, Van Nostrand Reinhold Company London, New York, pp 81-100.
[10] Kocherginsky, N.M., Yang, Q, Seelama, Lalitha, 2007. *Sep.and Purif.Tech.*,**53,** 171–177
[11] Beck M. T. 1990 *Chemistry Complex Equilibria.*, Ellis Horwood Limited, New York, , pp 174-179
[12] Rydberg 1992 *Priciple and Practices of Solvent Extraction,* Marcel D., New York, pp. 393-412

IOP Conf. Series: Materials Science and Engineering **205** (2017) 012011 doi:10.1088/1757-899X/205/1/012011

Cause Analysis on the Void under Slabs of Cement Concrete Pavement

Li Wen, Guo Xin Zhu and Baozhu

Communication and Transportation School of Changsha University of Science and Technology, China
617303487@qq.com

Abstract. This paper made a systematic analysis on the influence of the construction, environment, water and loads on the void beneath road slabs, and also introduced the formation process of structural void and pumping void, and summarizes the deep reasons for the bottom of the cement concrete pavement. Based on the analysis above, this paper has found out the evolution law of the void under slabs which claimed that the void usually appeared in the slab corners and then the cross joint, resulting void in the four sides with the void area under the front slab larger than the rear one.

1. Introduction
With the rapid development of the traffic construction in china, there is also a rapid increase in the mileage in highway open to traffic, along with a heavier maintenance task.
The pavement built in the early days were concrete pavement mainly. Affected by the repeat vehicle loads and environmental factors, there was an objective existence of void between some certain parts of the concrete slabs and the interlayers. The void under slabs has been one of the most significant problems for the concrete pavements. Once a broken slab, a kind of structure damage appeared, the maintenance would be quite difficult, not only including a high cost, but also large time consumption. Therefore, the discovery of the cause and evolution laws of void is of great practical and theoretical significance for the treatment of the broken slabs in rigid pavement.

2. Causes and Formation Mechanism of void Under Slabs

2.1. Causes of Void under Slabs
During the operation period of the cement concrete pavement, plastic deformation appeared on the basic level under the repeat influence of the traffic loads due to a rigidity of the base materials under the slabs weaker than the panel materials especially for the pavement with a the gravel as its base materials. When the wheel load acts on the road surface, the surface plate will be bent and deformed. As the deflection of the plate will cause a certain amount of plastic deformation, and the roadbed will remain part of the plastic deformation can not be restored.And with the accumulation of the plastic deformation, the void under the concrete slabs appeared. Meanwhile,, due to the uneven distribution of the temperature in the cement cement concrete pavement, it causes the buckling of the slab up or down, which accelerates the separation between the plate and the foundation.The void under slabs provided space for the intrusion of water. Cement concrete pavement has seams (and sometimes cracks) and free edges.During the rainy season, the rainwater infiltration from the gaps of the pavement formed ponding in the void area. Under the repeat squeezing action of the traffic load, the ponding together

with the fine particles in the base materials formed the mud which then caused pumping when splashed from the joints and cracks of the pavement. The production of pumping enlarged the void area under the slabs and then caused the breakage of the slabs with the interaction of overload vehicles. In conclusion, there are four main factors causing the void which are presented below:

2.1.1. Construction Factor

With the increase of service life, the increase of traffic volume, cement concrete substrate material will be a certain plastic deformation, the load away, this part of the plastic deformation can not be fully restored, will remain in the grassroots and soil for a long time, the formation of local Support conditions weakened or empty; the same time, because the material and construction of non-uniformity, the assessment of plastic deformation is more unpredictable.Due to the difference of the engineering geological conditions in the highway alignment areas, the longitudinal part usually adopted the design of secant lines. Most of them were fill subgrade (10m in height) and excavation subgrade (10m in depth), with some semi-filling and semi-excavating, no filling and shallow excavation subgrades, which made the construction control of the subgrade uniformity quite difficult. If the bright culvert and bridges were constructed in advance, the rolling equipment would not be able to get close to the culvert wall and the abutment back. As a result, the differential settlement of the filling area increased after the construction and also the void under the butt straps.

2.1.2. Influence of the Natural Factors

Changes in temperature and humidity will cause the pavement to warp, the force at this time is very unfavorable, under the action of the load is more prone to cracks. In order to avoid excessive temperature stress leaving the plate arch or broken, set up a variety of joints, but with the panel shrinkage, expansion, cracks are open, closed, these weak locations for the water immersion to create the conditions. This factor can not be avoided, but as long as dealing with other reasons, natural factors will not become the main reason for the void.Due to the infiltration of cement paste in the concrete during the process of cast-in-place, an integral material with a certain shear capacity formed between the slab and the subgrade. However, affected by the temperature, there would be an expansion of the slab which resulted in the decrease of the horizontal shear capacity. At the same time, there was a nonlinear temperature distribution in the slabs, which resulted in the upward or downward bend of the slab, accelerating the separation between the slab and subgrade. As a result, part of the cement concrete slabs lost the continuous touch with the slab thus causing the partially void under the slabs.

2.1.3. Scour of the Surface Water

The water inside the cement concrete pavement joints or cracks into the pavement, arrived at the site and at the base of weakening the void, under the effect of vehicle load, the basic materials of scour, deterioration of bottom supporting conditions, increasing the void area.Sometimes the subgrade was infiltrated due to the rainfall and other factors. When affected by the load effect, a high pressure flow would appear between the slab and the infiltrated subgrade, which would erode the subgrade surface. The whole process was the so-called scouring action. Once such action happened, part of the slab would lose the bearing of the subgrade thus causing the breakage and dislocation. At the same time, due to the dynamic load effect of the wheel, moisture will be carried out from the bottom of the joints and column points with the base material, which further aggravate the bottom of the slab.

2.1.4. Multiple Influence of the Vehicle Loads

Most of the transport vehicles seriously overloaded, overrun the road so that the actual number of standard load is much larger than the design value. Under the action of these vehicles, the pressure of the plate at the bottom of the pressure can be a few orders of magnitude larger than the standard car, which damage the road and its serious.When the vehicle load acted on the pavement, flexure and deformation would appear on the slab, causing certain deformation on the subgrade. The slab reverted after the vehicles left, however, there were some plastic deformation unable to be reverted in the subgrade. Although the amount of the rudimental plastic deformation each time was minimum, the accumulated amount after millions of times vehicle load actions were quite large. Therefore, a certain extent of void phenomena appeared in most carriageways after an operation of 3 to 5 years. Longitudinal and transverse joints were destroyed resulting in the penetrating cracks, which intensified the scour of surface water.

2.2. Formation Mechanism of the Void under Slabs

Cement concrete road surface layer and the grass-roots bonding as a whole of this state is impossible to maintain it, due to the concrete surface of the concrete layer of solidification, in the temperature and volume shrinkage, the surface layer and the grass between the The overall bonding state will soon change, resulting in the separation of the surface layer and the base layer to form a broken layer, and then in the role of rain and load, prone to empty.

The separation of the surface layer from the base layer will produce a small interlayer gap. Due to the brittle nature of the cement concrete pavement, the bearing capacity of the pavement will be reduced. The reason is that the separation as a form of destruction, which in the vicinity of the separation interface is accompanied by the destruction of the material, in the rain erosion, continuous vehicle load, etc. under the constant action of the separation of materials near the interface will be severely damaged, A crushing layer is formed between the topsheet and the base layer. The non-uniform expansion of the crust layer will aggravate the non-uniform voiding of the pavement in the transition layer between the surface layer and the base layer. In addition, under the constant impact of the vehicle load, the erosion of the crushing layer will be caused, damage.

AASHTO road test observations show that: in the cement concrete pavement and the location of the plate angle, there is a large number of cement mud due to the cement concrete slab under the foundation of the phenomenon.Based on large amount of survey data on the void phenomenon, the void under slabs in cement concrete pavement can be classified into two kinds. One is the structural void and the other is the pumping void.

2.2.1. Structural Void

Under the multiple factors of environmental factors (temperature change and humidity change) and vehicle load, the deflection of cement concrete panel will produce a certain amount of plastic deformation, the plate after the load away from the original state, and the roadbed remains this part of the plastic Deformation, the plate after the load away from the original state, although the load after each role of the roadbed residual plastic deformation is very small, but millions or even tens of millions of times after the cumulative amount is not small, the load on the road On different parts, the amount of deflection generated by the different corner of the corner, the edge of the board followed by the bottom of the plate under the smallest. In this way, the bottom of the plate and the edge of the board at the edge of the formation of the bottom of the roadbed between the top of the empty. Environmental factors caused by the accumulation of voids and plastic deformation are called structural voids.

2.2.2. Pumping Void

The rainwater infiltrated into the lower space of the slabs through the joins and the margin of the slabs, which formed pressured water under the multiple influence of the vehicle loads flowing in high speed along the joints and margins of the slabs. When the front wheel of the vehicle passes, the edge or corner of the rear panel is bent downwards, squeezing the water in the hollow area, and then the wheel is pushed to the rear. Driving speed is fast, the stagnant water is squeezed by the flow velocity is very high before and after the roadbed material in the high-speed water flow is easy to be washed, and stagnant water has become a fine mud containing mud.With the continuous loss of the fine particles, void formed under the slabs. This kind of void caused by pumping is called the pumping void.

3. The Evolution Law of the Void under Slabs

Under the condition of the positive temperature gradient, which means the temperature in the top surface is higher than the temperature in the bottom of the slabs, the void usually appeared in the middle lower space of the slab. While in the condition of negative temperature gradient when the temperature in the top surface was lower than the temperature in the bottom of the slabs, the void appears in the corners and margins of the slabs. Under the effect of vehicle load, the pressure on the corners and margins was larger due to a larger deflection in those areas. Plastic deformation appeared when the pressure was larger than the yield limit of those materials. The more times the vehicle passed the pavement, the larger the structural void space under the slabs was. When the loading acted on different parts of the slabs, the amount of deflection was also different, with the deflection amount in the corners larger than in the margins, and the middle parts of the slabs enduring the least defection

amount. Because the defection amount differs in different parts of the slab, the accumulated amount of plastic deformation was also different, with the amount in the corners larger than in the margins and the middle parts of the slabs having the minimum amount. Therefore, void between the slab bottoms and the subgrade surface appeared in the corners and margins of the slabs. The void means the slab has lost partial support which caused an evident increase of the flexural-tensile stress in the slabs inevitably. The research showed that the structural void under the slabs caused by Temperature warping and/or the accumulation of subgrade plastic deformation was relatively less both in degree and magnitude. Thus it had small effect on the increase of stress deflection and also a little influence on the enlargement of the void areas under the slabs. However, if the rainwater infiltrated into the structural void areas and the subgrade materials were easy to be washed away, thus, under the multiple influence of the vehicle loads, bumping was produced with fine particle being brought out continuously. The original void area would become larger and the stress deflection of the slabs increased at the same time, which resulted in the breakage of the slabs. Meanwhile, when the vehicle moves forward, the mud with fine particles jets in high speed to the void area of the rear slabs, leaving some fine clay deposited in the void area, thus elevating the rear slab and forming the vertical dispersion between the two joint unbroken slabs. This kind of problem is called dislocation which intensified the unevenness of the pavement. The structural void under the slabs was the cause of the pumping void, while the pumping void develops the structural void. The alternative or the combined action of the two kinds void promoted the evolution of the void under slabs.
Based on the field observation and measurement, when the vehicle passed a joint areas, the front and rear parts of the slab on the pavement were of great difference in the void scale, with the scour in the front and deposition in the rear, which means when the vehicle passed the corner of the first slab(the rear slab), due to the structural void area, part of the saturated water and sub-saturated water under the slabs expelled from the pumping channels under the effects of a relatively low pressure difference. When the vehicle passed through the front slab corners, the rear slab sprang back and a negative pressure produced under the slab. Under the effect of transient compression, the water under the front slab scoured the subgrade materials in the opposite direction at a high speed due to the mutual effects of a relatively high pressure difference and the pumping effects of the rear slab, bringing find particles back to the void area under the rear slab. A part of the fine particles deposited in the rear slab while others jetted out through the pumping channels (breathing joints and other cracks with poor tightness or the aged ones), resulting in a slow expansion of the void under the rear slabs due to the fine particles deposition and a continuous enlargement in the void area under the front slabs due to the water scour. Figure 1 is the general schema of the void under slabs.

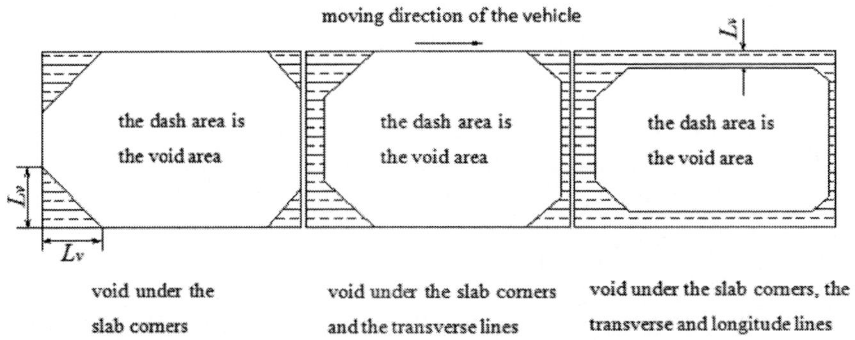

Figure 1. general schema of the void under a single slab

The evolution law of the void under the slabs usually starts from the corners and then to the transverse joint, forming a four sides void in the end. The void area under the front slab is larger than that in the rear slab based on the direction of vehicle movement.

IOP Conf. Series: Materials Science and Engineering **205** (2017) 012011 doi:10.1088/1757-899X/205/1/012011

4. Conclusion

(1) The void under the butt strap caused by the differential settlement during the construction, the nonlinear environmental temperature distribution, scour of the surface water and the repeat influence of the vehicle loads were the main reasons of the formation of the void under slabs of cement concrete pavements.

(2) The evolution law of the void under the slabs usually started from the corners and then to the transverse joint, forming a four sides void in the end. The void area under the front slab was larger than that in the rear slab based on the direction of vehicle movement.

5. References

[1] Liu Hai. Research on the Technology of Adopting Modified Emulsified Asphalt Cement Based Composite Material to Treat Void under Slabs of Cement Concrete Pavement [D].Changan University, 2013.

[2] Gao Shunping. Analysis on the Operation Conditions of Cement Concrete Pavement with Void under Slabs and Research on the Evaluation Technology of Void[D].Kunming University of Science and Technology,2014

[3] Zhang Yinchao. Research on Taking Rubber Powder Cement Mortar as the Grouting Agent for Void under Slabs of Cement Concrete Pavement [D].Hebei University of Technology, 2014.

[4] Zhang Jizhou. Cause Analysis on the Typical Diseases of JRCP Pavement of Lianyi Highway and Research on the Maintenance Strategies [D].Changsha University of Science and Technology, 2013.

[5] Wei Guojie, Feng Minglin. Cause Analysis and Treatment of the Void under Slabs of Cement Concrete Pavement [J].Science and Technology Information, 2013, (35):227.

[6] Tang Shanlin. Research on Typical Types of Diseases and Cause Mechanism of Cement Concrete Pavement in Cold Regions [D].Chongqing Jiaotong University, 2014

[7] Zhang Guoxiong. Research on the Void Mechanism and Fatigue Damage of Concrete Pavement Affected by the Dynamic Water pressure under the slab [D].Changsha University of Science and Technology, 2012.

[8] Zhang Xiaoran. Response Analysis on the void under the slab of cement concrete pavement affected by dynamic loads [D].Hebei University of Technology, 2015.

Novel Injectable Calcium Phosphate Bone Cement from Wet Chemical Precipitation Method

S Hablee[1], I Sopyan[1], M Mel[2], H M Salleh[2], M M Rahman[3] and R Singh[4]

[1]Department of Manufacturing and Materials Engineering, Kulliyyah of Engineering, International Islamic University Malaysia, PO Box 10, 50728 Kuala Lumpur, Malaysia.
[2]Department of Biotechnology Engineering, Kulliyyah of Engineering, International Islamic University Malaysia, PO Box 10, 50728 Kuala Lumpur, Malaysia.
[3]Department of Mechanical Engineering, College of Engineering, Universiti Tenaga Nasional, Jalan IKRAM-UNITEN, 43000 Kajang, Selangor Darul Ehsan, Malaysia.
[4]Department of Mechanical Engineering, Faculty of Engineering, University of Malaya, 50603 Kuala Lumpur, Malaysia.
sopyan@iium.edu.my

Abstract. Calcium phosphate cement has been prepared via chemical precipitation method for injectable bone filling materials. Calcium hydroxide, $Ca(OH)_2$, and diammonium hydrogen phosphate, $(NH_4)_2HPO_4$, were used as calcium and phosphorus precursors respectively. The synthesized powder was mixed with water at different powder-to-liquid (P/L) ratios, which was adjusted at 0.8, 0.9, 1.0, 1.1 and 1.2. The influence of P/L ratio on the injectability, setting time and mechanical strength of calcium phosphate cement paste has been evaluated. The synthesized powder appeared as purely hydroxyapatite with nanosized and agglomerated spherical particles. All cement pastes show excellent injectability except for the paste with P/L ratio 1.2. Calcium phosphate cement with P/L ratio 1.1 shows the ideal cement for bone filler application with good injectability, the initial and final setting times of 30 min and 160 min, and the compression strength of 2.47 MPa. The result indicated that the newly developed calcium phosphate cement is physically suitable for bone filler application. This paper presents our investigation on the effect of P/L ratio on the handling and mechanical properties of calcium phosphate cement prepared via wet chemical precipitation method.

1. Introduction

Calcium phosphate cement (CPC) is clinically accepted for bone cement application. This is attributed to their excellent biological properties, injectability and ability to harden *in vivo* [1,2]. CPC can be moulded and shaped to fill the defect sites by using injection needle. The commercial CPC is made of tetracalcium phosphate (TTCP) and α–tricalcium phosphate (α-TCP) which will transform to hydroxyapatite (HA) upon contact with water, forming apatite cements [3].
CPC can be synthesized via dry methods (i.e. solid-state and mechanochemical synthesis) and wet methods (i.e. chemical precipitation, hydrolysis, sol-gel, hydrothermal, emulsion and sonochemical methods) [4]. Wet chemical methods is the most common method used in preparation of HA due to their ability to control the morphology and mean size of CPC as well as its pore size by adjusting the reaction parameters compared to dry methods [4,5]. The synthesis of CPC via wet chemical precipitation method is the most favorable route due to the ease in experimental procedure, low synthesis temperature, high homogeneous purity products, low cost as well as high yield [4-10].

In this work, novel calcium phosphate cement has been successfully synthesized through wet chemical precipitation method. The synthesized powder was mixed with water at various powder-to-liquid (P/L) ratios. The present work aims at investigating the effect of P/L ratio on the mechanical and handling (i.e. injectability, setting time) properties of CPC, and an optimized condition was defined at which the obtained cement is physically suitable for bone filling materials.

2. Materials and Methods

2.1. Synthesis of powder

Calcium hydroxide ($Ca(OH)_2$) was supplied by PC Laboratory Reagent and diammonium hydrogen phosphate ($(NH_4)_2HPO_4$) by Friendemann Schmidt Chemical, Germany. CPC was synthesized based on the following reaction:

$$5Ca(OH)_2 + 3(NH_4)_2HPO_4 \rightarrow Ca_5(PO_4)_3OH + 6NH_4OH + 3H_2O \quad (1)$$

The HA powders were prepared via modified wet chemical precipitation method, using calcium hydroxide, $Ca(OH)_2$, and diammonium hydrogen phosphate, $(NH_4)_2HPO_4$, as the calcium and phosphorus precursors respectively, with distilled water as the solvent. Calcium-to-phosphorus (Ca/P) ratio was 1.67. Calcium and phosphorus solution was prepared separately, by dissolving each precursor in distilled water. Then, calcium solution was added drop wise into the phosphorus solution. The pH of the mixture was adjusted at pH 11 by adding 25% ammonia solution. The mixture was then refluxed at 90°C. The precipitate formed was left for overnight (18 hours) aging at room temperature. This was followed by washing with distilled water and filtering. The filtered precipitate was dried overnight (18 hours) in an oven at 85°C and then crushed. Figure 1 summarizes the wet chemical precipitation method.

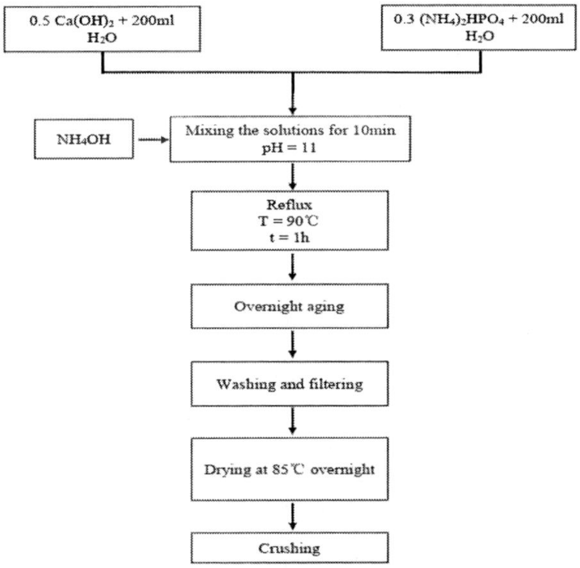

Figure 1. Flowchart of the wet-chemical precipitation synthesis.

2.2. Preparation of CPC

CPC was prepared by mixing powder and liquid phase for two minutes. The as-synthesized powder was used as the powder phase, and liquid phase is distilled water. P/L ratios were varied at 0.8, 0.9, 1.0, 1.1 and 1.2. The pastes were the put into a mould for further characterizations and properties evaluation.

2.3. Powder characterization

The as-synthesized powder was characterized for morphology and phase analysis. X-ray Diffraction (XRD) was used for phase analysis using an Empyrean PANalytical XRD system was operated by using CuKα radiation ($\lambda = 1.5406$ Å). Scan speed of 2° per minute and a step size of 0.02° over the 2θ range of 20-60° were applied. Field Emission Scanning Electron Micsroscopy (FESEM) JEOL JSM 6700F was employed for morphological evaluation.

2.4. Injectability test

The injectability test was done using Lloyd LR 10 K+ Universal Testing Machine under compression mode. This is called extrusion method, in which the cement paste was extruded out of non-needle 5ml polyethylene syringe. The extrusion was done by applying a crosshead speed of 50 mm/min and a maximum load of 300 N.
The result of this extrusion method was recorded as the evolution of extrusion force (N) against the extrusion time (sec) and plotted in y and x axes respectively. The injectability tests were evaluated for CPC with different P/L ratios.

2.5. Setting time

The determination of the setting time of CPC was done using Gillmore needle method. After mixing the powder and liquid phases for two min, the paste was put into a polyethylene mould of 8 mm diameter x 12 mm length dimension. This work determined the setting time for CPC with different P/L ratios.
The initial and final setting times were determined by applying the thick and thin needle respectively in vertical position until no visible indentation was observed. Using Gillmore needle apparatus, initial setting time is the time when CPC able to resist the small fixed pressure applied by a thick Gillmore needle, while final setting time is the time when CPC able to resist the high fixed pressure applied by thin Gillmore needle [1,2].

2.6. Mechanical test

The measurement of compression strength was done using Lloyd LR 10 K+ Universal Testing Machine. The cement pastes were molded in Teflon moulds and left to dry for 48 hours, followed by compression tests on the 10 mm diameter x 15 mm length specimen at 1 mm/s crosshead rate.

3. Results and Discussion

Numerous studies on calcium phosphate cement development have used more complex synthesis routes and the synthesis time is relatively long [6, 10-15]. The modified wet chemical precipitation method in the present work is simpler and shortened the synthesis time.
Figure 2 shows the XRD patterns of the synthesized powder. The peaks present in the figure attributed to the standard data of HA (card no. 024-0033) as the main phase. The sharp and clear peaks confirmed the phase purity and crystallinity degree of the synthesized powder.
The FESEM image of synthesized HA powder is presented in Figure 3. The micrograph shows the agglomeration of spherical HA particles. The picture also is an evidence of the formation of nano size powder. This is in agreements with previous studies done by Yousefi *et al.* [11] and Dhanalakshmi *et al.* [12].

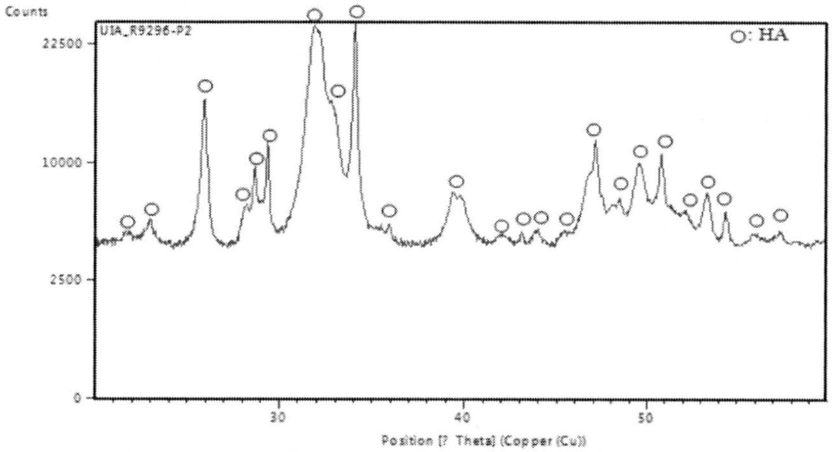

Figure 2. XRD pattern of the synthesized HA powder.

Figure 3. FESEM image of as-synthesized HA powder.

CPC was prepared by mixing the powder and distilled water for two min at different P/L ratios, which are varied from 0.8, 0.9, 1.0, 1.1 and 1.2. Injectability test of CPC as shown in Figure 4 indicated that the increase in powder content increase the extrusion load and injectability become poorer. This happen due to the increase in powder content which increase viscosity of the paste and become hard to be injected. CPC with P/L ratio 1.2 has become unworkable and the paste cannot be extruded out of the syringe.

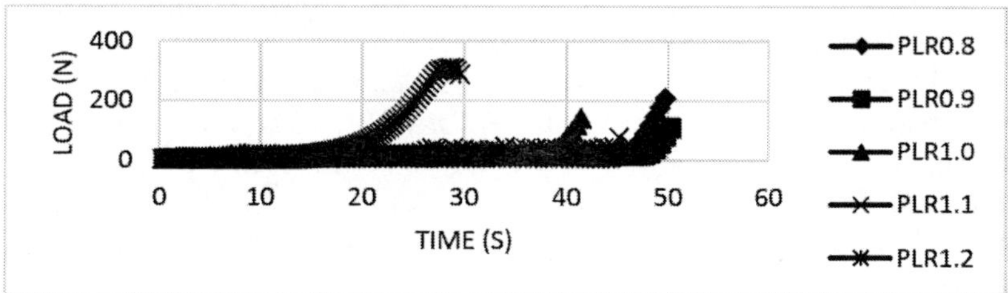

Figure 4. Injectability of CPC at different P/L ratios.

From Figure 5, the increase in P/L ratio shorten the setting time of CPC. This is due to the increase in powder content in the cement paste. P/L ratio 1.1 shows the optimum workable setting times of CPC, with initial setting time of 30 min and final setting time of 160 min. When CPC was prepared using P/L ratio 1.2, the paste was not workable with both initial and final setting times are 3 minutes. This is in a good agreements with the injectability result as the paste with P/L ratio 1.2 also shows no injectability.

The strength of CPC was determined for different P/L ratios. The result for compression test shown in Figure 6. The average strength of CPC ranged from 1.23 MPa to 2.47 MPa. CPC with P/L ratio 1.1 shows the highest strength with 2.47 MPa. It seems the entangled networks of apatite crystals become denser with the increase in powder content [1]. But, P/L ratio 1.2 CPC is an exception because it only forms a wet powder and hence dissolution and reprecipitation might not induced.

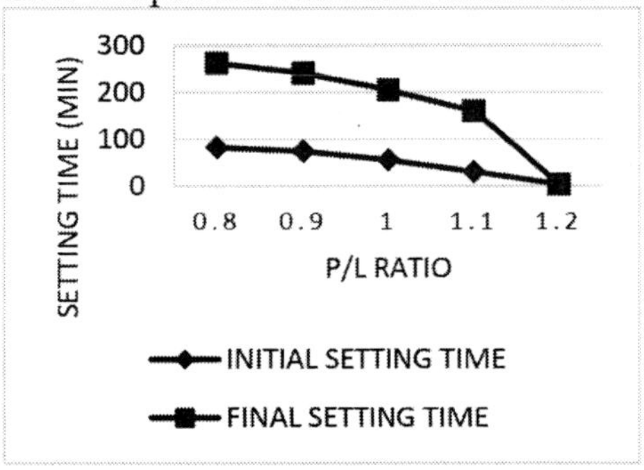

Figure 5. Setting time of CPC with different P/L ratios.

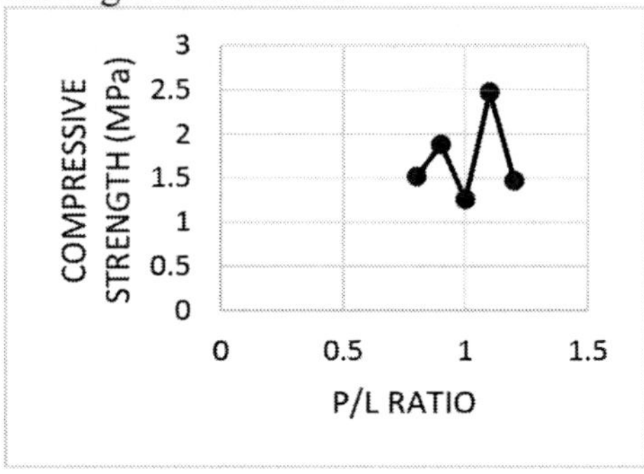

Figure 6. Compressive strength of CPC with different P/L ratios.

4. Conclusion

A novel wet chemical precipitation method has successfully produced a pure nano scale HA by using calcium hydroxide and diammonium hydrogen phosphate. The determination of injectability, setting time and compressive strength have been successfully done to elucidate the effect of P/L ratio on CPC. The results show that the increase in P/L ratio shortened the setting time and improved compressive strength, but reduced injectability. The synthesized CPC has proven its ability as injectable bone filling materials with optimized condition at P/L ratio 1.1.

References

[1] Zhang J, Liu W, Schnitzler V, Tancret F and Bouler J 2014 *Acta Biomaterialia* **10** 1035
[2] Hablee S, Razali N N, Alqap S F and Sopyan I 2016 *Recent Patents on Materials Science* **9(2)** 72
[3] Alqap A S F, Sopyan I, Husni M and Athirah N 2012 *Applied Mechanics and Materials* **110-116** 8
[4] Sadat-Shojai M, Khorasani M, Dinpanah-Khoshdargi E and Jamshidi A 2013 *Acta Biomaterialia* **9** 7591
[5] Mohammad N F, Othman R and Fei Y 2014 *Rev. Adv. Mater. Sci* **38** 138
[6] Monmaturapoj N and Yatongchai C 2010 *Journals of Metals, Materials and Minerals* **20(2)** 53
[7] Nayak A K 2010 *Int. J. ChemTech Res.* **2(2)** 903
[8] Pataquiva-Mateus A Y, Ferraz M P and Monteiro F J 2013 *Revista Mutis* **3(2)** 43
[9] Cox S 2012 *Lucideon (formerly Ceram)* pp 1-10
[10] Monmaturapoj N 2008 *Journals of Metals, Materials and Minerals* **18(1)** 15
[11] Yousefi K, Zebarjad S M and Khaki J V 2015 *J. Sol-Gel Sci. Technol* **76** 592
[12] Dhanalakshmi C P, Vijayalakshmi L and Narayan V *2012 International Journal of Physical Sciences* **7(13)** 2093
[13] Ungureanu D N, Angelescu N, Ion R M, Stoian E V and Rizescu C Z 2011 *Bojkovic, Z.; Kacprzyk, J.; Mastorakis, N.* **20** 296
[14] Wang P, Li C, Gong H, Jiang X, Wang H and Li K 2010 *Powder Technology* **203** 315
[15] Eslami H, Solati-Hashjin M, Tahriri M and Bakhshi F 2010 *Materials Science-Poland* **28(1)** 5.

Acknowledgement

This work was financially supported by Ministry of Education (MOE) through Fundamental Research Grant Scheme (FRGS) with the project ID of FRGS15-246-0487.

IOP Conf. Series: Materials Science and Engineering **205** (2017) 012013 doi:10.1088/1757-899X/205/1/012013

Effect of Poly(Vinyl Alcohol) Addition on the Properties of Hydrothermal Derived Calcium Phosphate Cement for Bone Filling Materials

N N Razali[1], **I Sopyan**[1], **M Mel**[2], **H M Salleh**[2], **M M Rahman**[3] **and R Singh**[4]

[1]Department of Manufacturing and Materials Engineering Department, Faculty of Engineering, International Islamic University Malaysia, PO Box 10, 50728 Kuala Lumpur, Malaysia.
[2]Department of Biotechnology Engineering, Faculty of Engineering, International Islamic University Malaysia, 50728 Kuala Lumpur, Malaysia.
[3]Department of Mechanical Engineering, Faculty of Engineering, Universiti Tenaga Nasional, 43000 Kajang, Selangor.
[4]Department of Mechanical Engineering, Faculty of Engineering, University of Malaya, 50603 Kuala Lumpur.
sopyan@iium.edu.my

Abstract. The effect of addition of poly(vinyl alcohol) on hydrothermal derived calcium phosphate cement has been studied. The precursors used to prepare the cement were calcium oxide (CaO) and ammonium dihydrogen phosphate ($NH_4H_2PO_4$); the reaction was conducted in water at 80-100°C. To improve properties of CPC, poly(vinyl alcohol) (PVA) of 1wt% and 2wt% was added to the liquid phase of CPC and the results were compared to CPC without PVA addition. The addition of PVA was proved to bring remarkable effects on cohesion, setting time and mechanical strength of CPC which make it suitable physically for injectable bone filler applications.

1. Introduction

Calcium phosphate cements (CPC) has shown great interest as one of the materials for bone graft applications because of excellent biological behavior, non-toxicity, biocompatibility, and osteoconductivity[1]. CPC is generally produced by chemical reaction between two phases which are powder phase and liquid phase[2,3]. The preparation of cement using hydrothermal method is most favorable because its ability to produce highly crystalline cement with high yield, controlled morphology and stoichiometric Ca/P ratio[5].
Nevertheless, there are certain drawbacks that need to be considered which gives limitation to its clinical applications[2,6]. CPC has shown poor injectability and in certain cases, the phase separation between powder and liquid might occur[7]. The cements also have tendency to disintegrate after injected and contact with blood or other physiological solutions in body because of weak cohesion[7].
Numerous works have been reported to increase integrity of CPC by adding polymeric binder such as poly(vinyl alcohol) (PVA), poly(lactic-co-glycolic acid) (PLGA), chitosan fibers, polyamides and others to either powder or liquid phase of CPC paste with the aim to improve their handling and mechanical properties[6]. Among these polymers, PVA is of great interest from the viewpoint of improving cohesion and setting time of CPC. PVA is one of the hydrophilic biocompatible polymer that was widely used in biomedical applications. The high hydrophilicity and easy swelling when

absorbing water have made PVA as one of the materials suitable to maintain stability of the CPC during storage[8]. In this report, the effect of PVA addition on the handling and mechanical properties of hydrothermal derived CPC will be presented.

2. Materials and Methods

2.1 Synthesis of powder.
The calcium phosphate cement (CPC) was prepared using a hydrothermal method as we reported elsewhere[9]. The chemical reaction to synthesize CPC powder is as follows:

$$5Ca(OH)_2 + 3(NH_4)_2HPO_4 \rightarrow Ca_5(PO_4)_3OH + 3NH_4OH + H_2O \quad (1)$$

The composition of CPC consists of calcium oxide (CaO) and ammonium dihydrogen phosphate (ADP), $NH_4H_2PO_4$. The stoichiometric composition of hydroxyapatite (HA) produced is referred to 0mol% excess of CaO. The synthesized HA was then dried overnight at 80°C until the dry powder was produced. To make CPC paste, the dried powder (p) and distilled water (w) were mixed homogenously for 3min at the ratio of (3:2) p/w mass-ratio for preparation of filler materials. The evaluation of the effect of PVA addition on the CPC paste was done by addition of 1 and 2wt% of PVA to the filler materials. The paste was then filled into the 10mm diameter and 15mm height PTFE (Teflon) mold by hand-spatulation. Figure 1 shows the flowchart of the method to produce CPC.

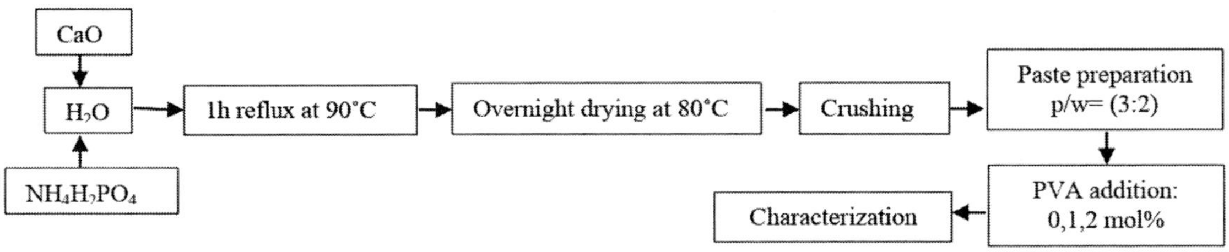

Figure 1. Flowchart of the methods to produce CPC with PVA addition.

2.2 Physical and Mechanical characterization
The dried powder obtained was subjected to physical and mechanical characterization. XRD was used for crystal structure and phase analysis using Empyrean, PANalytical XRD system. The scanning done at scan speed of 2° per minute and a step size of 0.02° over the 2θ range of 20-60°. The diffraction patterns obtained from the XRD analysis was compared to powder diffraction files from JCPDS database to determine the phase of HA. Morphological analysis on the powder was done using JEOL, JSM 6700F.

2.3 Injectability test
The injectability test was done by extrusion of 5 ml non-needle syringe with 2.68mm inner diameter of canula filled with paste as the working sample. The force applied on the syringe ejector spindle was determined by recording the evolution of the extrusion force (N) against the extrusion time (sec) using of a Lloyd universal testing machine, LR 10 K+ model, at a crosshead speed of 50mm/min and a maximum load of 300N. The injectability results for CPC samples with addition of 1 and 2wt% of PVA were evaluated.

2.4 Setting time test
To identify the setting time, Gillmore method was used to measure (3:2) p/w ratio of CPC with the addition of PVA. After homogenously mixed, the CPC paste was moulded into PTFE (Teflon) mold. The initial and final setting times were determined by holding the light and heavy needles in vertical position and applied to the cement surface until no visible indentation formed.

2.5 Mechanical test

The compression test was done using the Lloyd LR 10 K+ Universal Testing Machine on 10mm diameter x 15mm length specimens at 1mm/min crosshead rate for all samples of CPC.

3. Results and Discussion

The addition of 1 and 2wt% of PVA has shown remarkable effects on injectability, setting time, morphology, cohesiveness, density and mechanical strength of calcium phosphate cement.

The addition of 1 and 2wt% of PVA to CPC was able to extrude all paste from the syringe as shown in Figure 2. CPC without PVA shows good injectability and started to increase load for paste extrusion at 40s. However, the paste has shortened the extrusion time to 37s and 33s for 1 and 2wt% PVA added to CPC, respectively, due to absorption of water molecules by PVA which leads to more viscous and poor injectability of CPC paste. Additionally, Figure 3(a) and (b) represents both CPC without PVA and CPC with 1wt% PVA, able to extrude all the paste from the syringe. Meanwhile, there are little paste still remained in the syringe as shown in Figure 3(c) for CPC with 2wt% PVA. Overall, the increase in the concentration of PVA decreases the injectability.

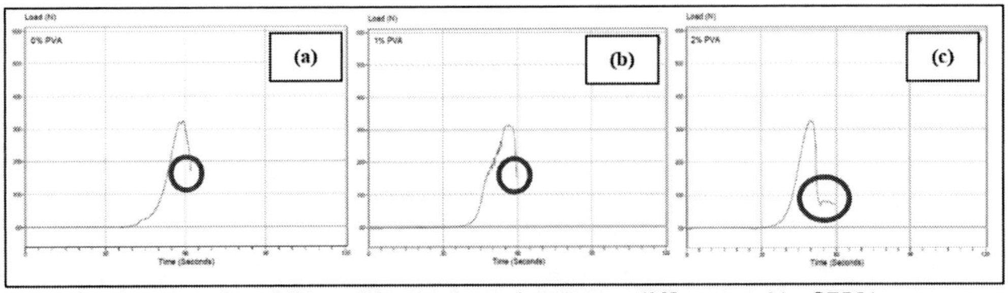

Figure 2. Injectability of (3:2) p/w ratio CPC at different wt% of PVA: (a) 0%, (b) 1% and (c) 2%.

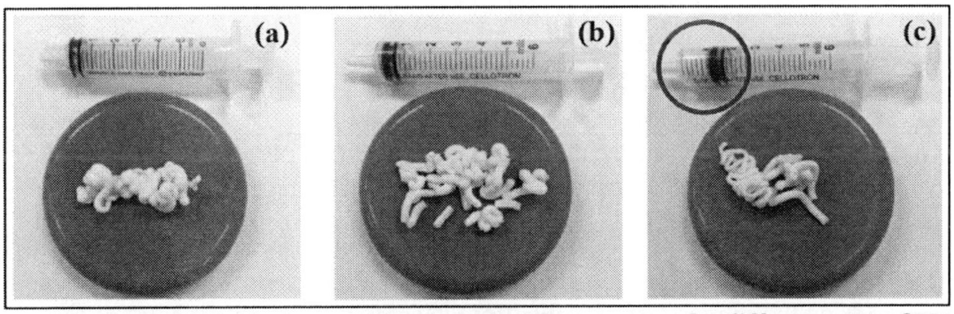

Figure 3. Moldable forms of calcium phosphate cement for different wt% of PVA: (a) 0%, (b) 1% and (c) 2%.

Figure 4 shows the initial and final setting time of CPC as the function of PVA concentration. The initial setting time of CPC without PVA is 11min, and decreased to 6 and 3min respectively, when PVA concentration of 1 and 2wt% was added. While, the final setting time of non-PVA CPC paste was 87min, and it decreased to 56 and 38min when PVA was added to the concentration of 1 and 2wt%, respectively. Rajzer *et. al* reported that addition of PVA with the presence of glycerol to CPC increases the setting time, might be caused the high viscosity of glycerol [8].

The presence of hydroxyapatite was marked by a strong x-ray peak at (31.733; 2θ). The clear and sharp peak shown in Figure 5 proves that the phase purity and crystallinity of the hydroxyapatite powder. The crystallinity of CPC was also mentioned previously in the study done by Wang *et. al* [10] and Yousefi *et al* [11]. CPC without PVA shows slightly higher peak of HA compared to CPC with PVA addition.

Figure 6 presents FESEM for fine powder of HA where nanosize particles of 100nm long are clearly shown in CPC without PVA. The powder also formed agglomerated particles of a needle like morphology. After addition of 1wt% PVA, fine powder of HA with nanosize particles of 100nm long still visible but there is more agglomeration of spherical size particles compared to CPC without PVA. The addition of 2wt% PVA to CPC gave different looks to microstructure of CPC where the powder particles have been covered by PVA particles, making the CPC particles invisible.

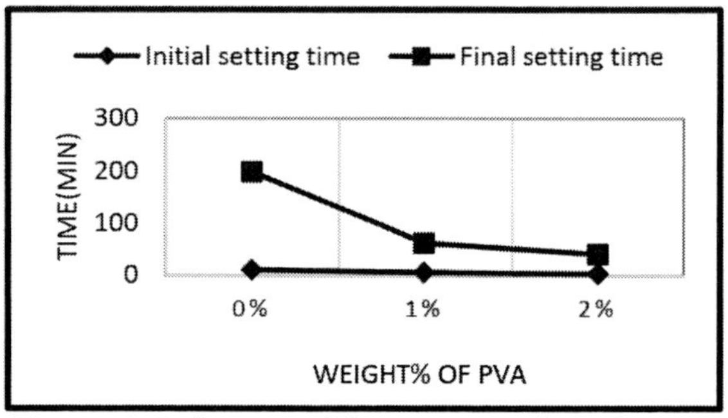

Figure 4. Influence of the concentration of PVA on the setting time of CPC.

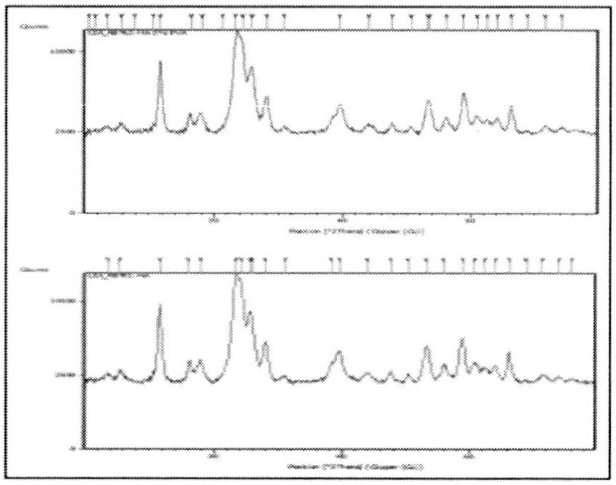

Figure 5. XRD peak of CPC without PVA(above) and CPC with the addition of 2wt%PVA(below).

Figure 6. Morphology of the nanosized particles of CPC without PVA (a), CPC with 1wt% PVA (b) and CPC with 2wt% PVA (c) tested using FESEM.

In general, the addition of PVA was very effective in improving the compression strength of the CPC. This test is beneficial to determine the strength of the materials and make sure it is able to withstand the load depends on the applications. Results from the compression strength test as shown in Figure 7, proves that the addition of 1wt% PVA increases the compression strength of CPC from 1.01 to 2.03Mpa and increases up to 2.24MPa after addition of 2wt% PVA. This shows that PVA addition has improved the mechanical strength of CPC.

Figure 8 shows the relative density of CPC with different wt% PVA added to CPC. After addition of 1 and 2 wt% of PVA to CPC, the density of cement decreased from 1.769 to 1.721 g/cm^3. Meanwhile, the relative density of CPC can be calculated as compared to density of hydroxyapatite that is 3.14 g/cm^3. From the calculations, the relative density of CPC without PVA, CPC with 1 wt% PVA and CPC with 2 wt% PVA are 43.66%, 44.42% and 45.19%, respectively.

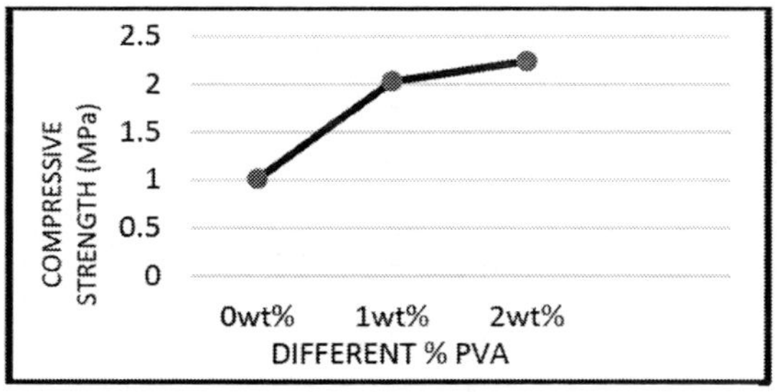

Figure 7. Compressive strength of CPC without PVA and CPC with addition of 1 and 2wt% PVA.

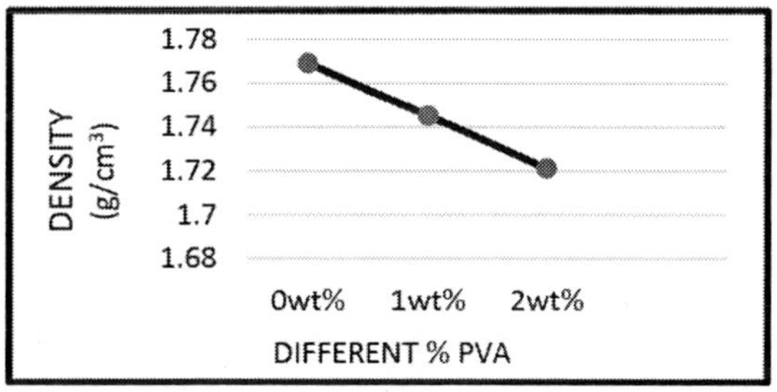

Figure 8. Density of CPC without PVA and CPC with addition of 1 and 2wt% PVA.

4. Conclusion

The effect of PVA addition on the physical properties of hydrothermal derived CPC has been investigated. The as-synthesized powder was purely hydroxyapatite (HA) and no tangible change observed in its crytallinity after PVA addition. The microstructure size increased with PVA concentration but still maintain their spherical shape and agglomerate. The results also showed that, the initial and the final setting times of CPC have been faster with 2wt% PVA addition from 8 to 3 min and 87 to 38 min, respectively. However, CPC with 2wt% PVA did not show injectability as excellent as pure CPC showed where some portion of the paste remained in syringe after extrusion. The extrusion time also decreased from 40s to 33s after the PVA addition proving a poorer injectability. CPC's compression strength increased from 1.01 to 2.24 MPa after addition of 2wt% PVA revealing much better ductility. In this study, PVA has acted as a suitable setting agent for the hydrothermal derived CPC, that helps to reduce setting time and improved compressive strength of the CPC cement. Therefore, the injectable CPC-PVA developed can be a promising material used in hard tissue filling applications.

Acknowledgement
This work was financially supported by Ministry of Education (MOE) through Fundamental Research Grant Scheme (FRGS) with the project ID of FRGS15-246-0487.

References

[1] Sariibrahimoglu K, Leeuwenburgh SCG, Wolke JGC and Yubao L 2012 *Journal of Biomedical Materials Research Part A* 712.
[2] Zhang J, Liu W, Gauthier O, Sourice S, Pilet P, Rethore G, Khairoun K, Bouler JM, Tancret F and Wiess P 2014 Acta Biomaterialia 1035.
[3] Zhang J, Liu W, Schnitzler VA, Tancret F and Bouler JM 2015 *Acta Biomaterilia* 1.
[4] Mohammadi M, Herasaki S and Ardakani M H 2012. Effect of adding nano-silicon carbide on mechanical properties and hydroxyapatite formation in calcium phopshate cements, *Iranian Conference on Biomedical Engineering* (Tehran) 2012.
[5] Nathanael AJ, Hong SI, Oh TH, Seo YH, Singh D and Han SS 2016 *Royal Society of Chemistry* 6
[6] Hablee S, Razali NN, Alqap ASF and Sopyan I 2016 *Recent Patents on Materials Science* **9** 72
[7] Liu W, Zhang J, Weiss P, Tancret F and Bouler JM 2013 *Acta Biomaterialia* 5740.
[8] Rajzer I, Castano O, Engel E and Planell JA 2010 Journal Material Science 2049.
[9] Sopyan I, Alqap ASF, Razali NN, Mel M and Suryanto, Development of Injectable Calcium Phosphate Cement for Bone Filling Materials, *8th Widyamata International Seminar on Sustainability* (Bandung) 2016
[10] Wang Q, Yan J, Yang J and Li B 2016 Nanomaterials promise better bone repair. *Mater. Today* http://dx.doi.org/10.1016/j.mattod.2015.12.003
[11] Yousefi KH, Zebarjad SM and Khaki JV 2015 *J. Sol-Gel Sci. Technol* **76** 592

IOP Conf. Series: Materials Science and Engineering **205** (2017) 012014 doi:10.1088/1757-899X/205/1/012014

Experimental and Numerical Analysis of Thermal and Thermomechanical Behavior of a Power Inductor Accompanied by a Reliability Study

Om Bendaou[1,2], C Gautrelet[1], A ELHAMI[1], M Agouzoul[2]

[1] INSA-Rouen, LOFIMS, France
[2]University Mohammed V, EMI, ERD3M, Rabat-MOROCCO

Abstract. In practice, when feeding the electrical and electronic components, they generate heat by Joule's effect, which will cause an increase in temperature and provoking mechanical deformations related to the expansion of the materials. It is therefore this aspect that is the subject of our study on the thermal and thermomechanical characterization of a ferrite power inductor by experimental and numerical channels. For this purpose, a finite element model was developed and calibrated by comparing its results to the experimental results obtained via thermal measurements and measurements by speckle interferometry system. A reliability analysis on the power inductor was performed numerically, based on the coupling between finite element model validated and the methods of Monte Carlo, FORM and SORM.

1. Introduction

The studied power inductor belongs to a dc-dc converter system. It is composed of a copper coil glued to a ferrite core. Ferrite was chosen in the composition of the power inductor because of the effectiveness of their magnetic and electric properties. However, the ferrite has a drawback of a mechanical nature, owing to its fragility and its sensitivity to temperature variations [1]. In fact, during operation of the power inductor, it will be exposed to a high amperage supply (up to 130A). This will create a temperature rise due to heat loss and will entail a failure mode and even cause rupture of the ferrite core (figure 1).

(a) Rupture of the Power inductor

(b) The power inductor

Figure 1. The power inductor studied

This article focuses on the study of the thermal and thermomechanical behavior of a power inductor, given its high sensitivity to thermal stresses which they are exposed. To do this, we developed a finite element model of a 3D power inductor with ANSYS, to view its thermal behavior, when the variation of the electric current of 0A to 50A and also calculate the mechanical displacement generated by the thermal stresses. To confirm and validate the numerical results, experimental tests were previously-

IOP Conf. Series: Materials Science and Engineering **205** (2017) 012014 doi:10.1088/1757-899X/205/1/012014

performed using thermal measurements and mechanical measurements by an electronic speckle pattern interferometry (ESPI).

Uncertainties in the physical properties of ferrite such as thermal conductivity or coefficient of thermal expansion can significantly affect the performance of the power inductor. It is for this reason that the purely deterministic analysis has been replaced by a probabilistic analysis that takes into account the uncertainties of the ferrite material properties. Indeed, a reliability study on power inductor was undertaken, with application of three reliability analysis methods namely: the Monte Carlo method and methods Form / Sorm to calculate the probability of failure [2].

2. Experimental measurements

To supply the thermomechanical numerical model, it is measured synchronously the growth in temperature at different points of the inductor and the displacements undergone by the ferrite core surfaces. It was decided not to work on a single inductor but on a complete dc-dc converter system so as not to block the possible deformations and be as close as possible to real conditions functioning. The figure 2 shows the fixation systems of the dc-dc for measurements on the side and on top of the inductor.

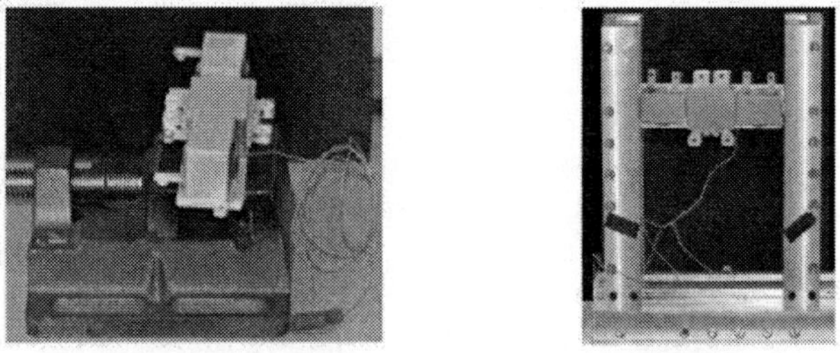

(a) Measurement on the side (b) Measurement on the top

Figure 2. Fixation systems of the DC-DC

To realize the thermal load in the structure, the inductor is fed by a current source. The coil of the inductor heats by Joule effect and communicates this heat to the ferrite core.

2.1. *Thermal* measurements

The temperature measurements are made by using six thermocouples affixed at several points of the inductor. Two thermocouples in top and bottom of the coil, three thermocouples in top,midle and bottom of the ferrite and the last to measure the ambiant temperature. Figure 3 describe these measurement points.

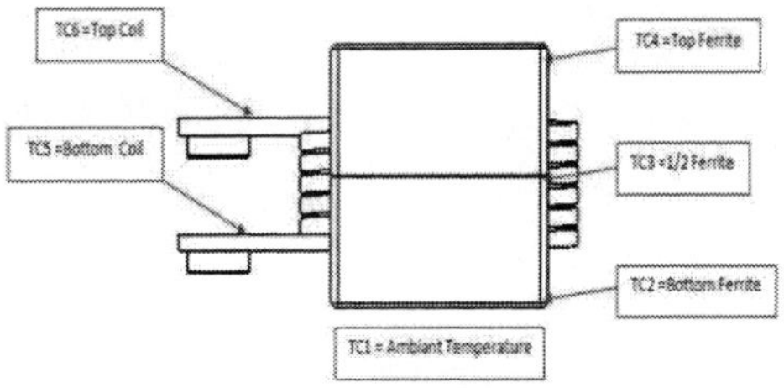

Figure 3. Position of the thermocouples

Measures acquisition is made by the module TC08 of Pico Technology. A Labview routine was developed for synchronization with image acquisition.

2.2. *Optical* measurements

In order to obtain the displacement of the ferrite core surfaces, a field measuring is required via an interferometry system composed by: a CCD camera; a Nd-YAG laser source of *100mW;* piezo actuator attached to a mirror; a computer system including a holographic processor [3-5].

IOP Conf. Series: Materials Science and Engineering **205** (2017) 012014 doi:10.1088/1757-899X/205/1/012014

The video signal from the camera is sent simultaneously to a video acquisition module Managed National Instrument by Labview which will allow to post process the raw images. The video capture card is synchronized by the RTSI bus with a general acquisition module, which allows to synchronize with temperature measurements. The real image (figure 4a) is given for information. The measures consists of acquiring raw images with specklegramme (figure 4b) and calculate the cards of phases folded (figure 4c). The cards of phases folded of reference calculated from the first 4 images is subtracted to the following phase maps [6].

(a) Real image (b) Specklegramme (c) Folded Phase

Figure 4. Overview 1

The phases folded being modulo *2π*, it's necessary to demodulate them to obtain the displacement. To achieve this, many phase demodulation algorithms can be used. The reference [7] is one such example. For this study, we will use one provided by the Fringe Analysis 4 of HOLO3 and the Matlab routines developed in the laboratory, in particular using [8]. Starting from the unfolded phase φ reduces to the surface studied and the wavelength of the laser λ illuminating the test structure, the displacement field off plan is given by:

$$d = \lambda / 4\pi \times \varphi \tag{1}$$

3. Numerical Model

Solving a problem by thermomechanical finite element method is performed in two steps: the first step is the resolution of the thermal problem. In order to the temperature field independently of the mechanical conditions. The second step is evaluate deformations produced by this temperature field.
The finite element model relating to the power inductor is composed of a coil (5.5 turns) glued to a ferrite core and the assembly is bonded to an aluminum plate (figure 5).

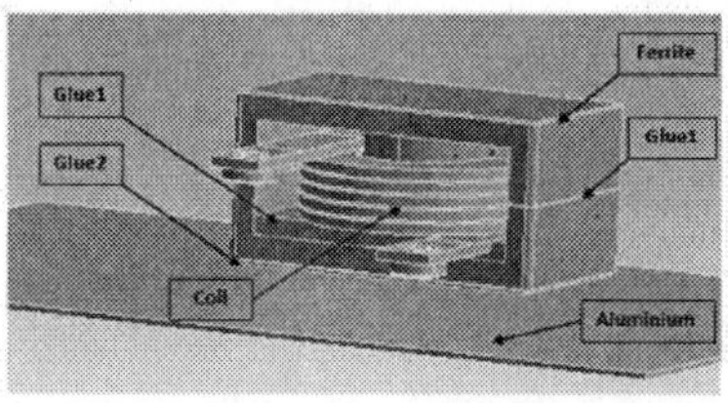

Figure 5. Power inductor modeled in ANSYS

Figure 6. Power inductor mesh

The geometrical properties of the ferrite core are shown in figure 7 and material properties in table 1.

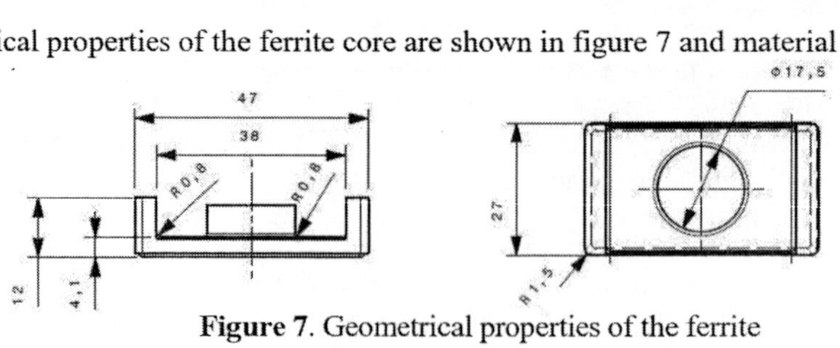

Figure 7. Geometrical properties of the ferrite

IOP Conf. Series: Materials Science and Engineering **205** (2017) 012014 doi:10.1088/1757-899X/205/1/012014

Table 1. Material properties of power inductor.

	Ferrite	Copper	Aluminum	Glue 1	Glue 2
Thermal conductivity (W/mK)	5	400	235	0.4	1.8
Coefficient of thermal expansion	12e-6	1.66e-5	2.3e-5	4.5e-5	4.5e-5
Young's modulus (Pa)	150	1.14e11	7.3e10	3.55e9	3e9
Poisson's ratio	0.28	0.355	0.33	0.3	0.3

The temperature field and the displacement out of plane were calculated on the side of the ferrite and on top for amperage of 50A. The DC resistance of the coil measured by an ohmmeter is of the order of $0.82\ m\Omega$. When applying a direct current, the heat losses are type RI^2.
The applied loads: Heat losses for 50A on the ferrite and the coil; Room Temperature *20°C.*

4. Presentation of the results

4.1. Thermal measurements
Following the electrical excitation sequence, is measured the evolution of the temperature (figure 8) at different locations of the inductor.

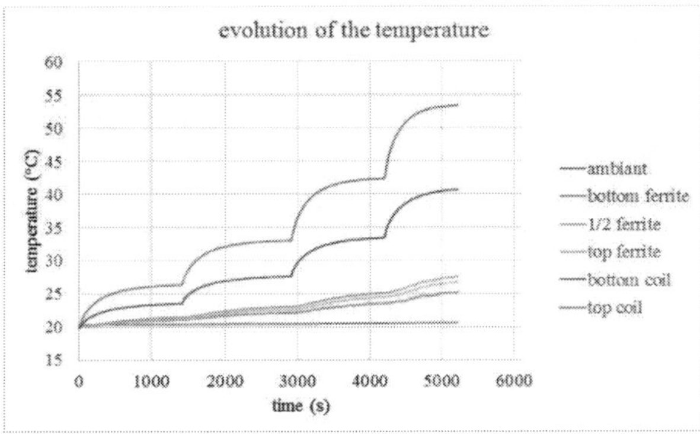

Figure 8. Evolution of the temperature in the inductor.

4.2. Optical measurements

4.2.1. Measurements on the side
The following figures 9 show the displacement of the surface, side view of the ferrite of the inductor. On this surface, the vertical profile half-base and the horizontal profile at mid-height are plotted and represented in parallel.

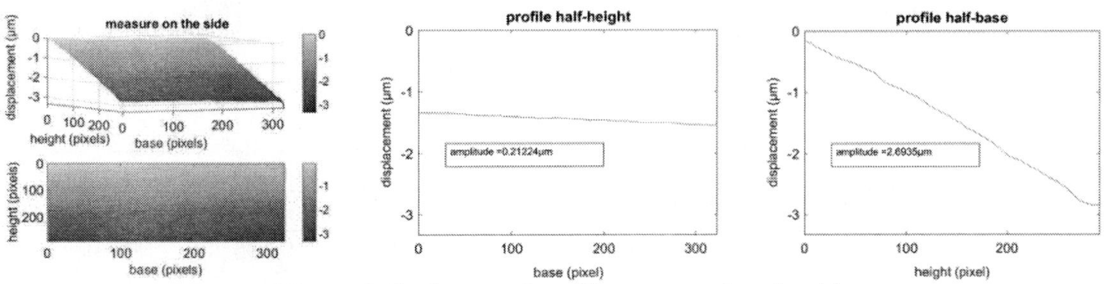

Figure 9. Surfaces and profiles measured on the side.

4.2.2. Measurement on top

The following figure 10 how the displacement of the surface, top view of the ferrite of the inductor. On this surface, the vertical profile half-base and the horizontal profile at mid-height are plotted and represented in parallel.

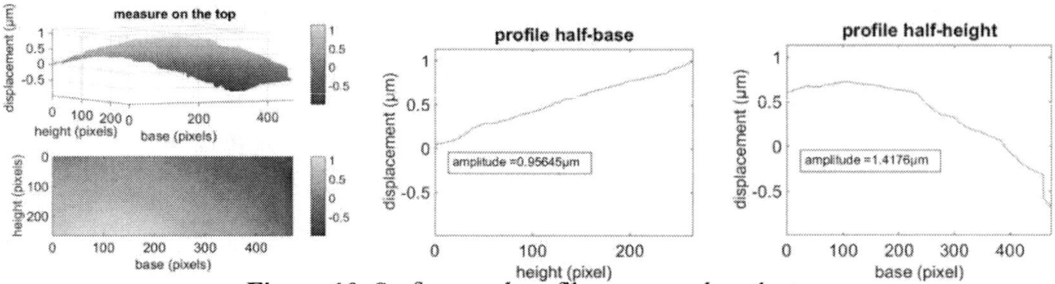

Figure 10. Surfaces and profiles measured on the top.

4.3. Numerical results compared with speckle interferometry measurements

We proceeded to a confrontation between the experimental and numerical results on the power inductor. We compared the thermal and thermomechanical results numerically and experimentally, while giving the relative error between these results.

- **Thermal results**

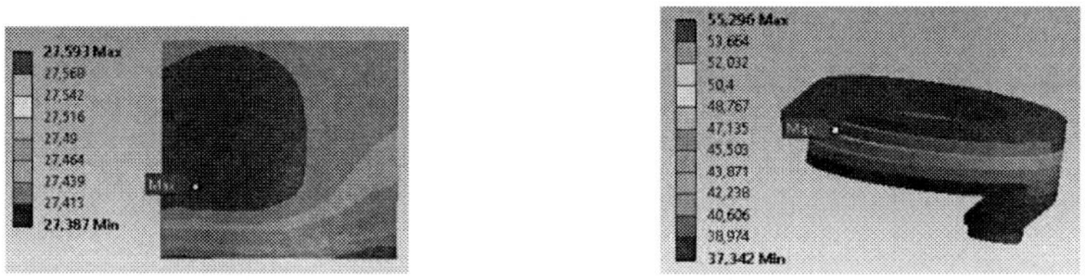

Figure 11. Temperature fields on the coil and on the side of ferrite

Table 2. Comparison of thermal results for 50A.

	Experimental results	FEM result
Ferrite	27°C	27°C
Coil	53°C	55°C

- **Thermomechanical results**

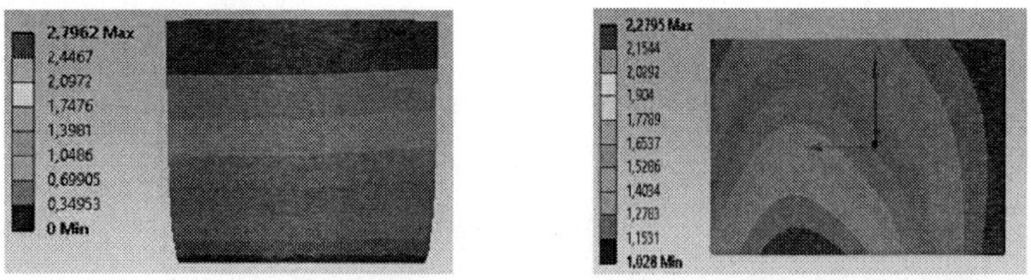

Figure 12. Displacement out of plane on side and top of the ferrite

Table 3. Comparison of thermomechanical results for 50A.

	Experimental results	FEM result	relative error
displacement amplitude out of plane on the side	2.693	2.796	3.8%
displacement amplitude out of plane on the top	1.417	0.135	11%

5. Results of the reliability study

A sensitivity study is required to select the most significant variables influencing on the system status. The other variables that play a minor role are supposed deterministic. The sensitivity study relating to power inductor, which concerned the material properties, allowed us to predict that the variation of CTE ferrite largely affects on the displacement of the side of the ferrite.
We estimated that the CTE of the ferrite follows a normal distribution with mean *1.2e-5* and standard deviation of *1.73e-7*. The failure is based on the fact that at *50A* the displacement on the side of the ferrite should not exceed 2.9945µm, while specifying that this value is taken in an arbitrary manner. The objective function can be written as follows:

$$G = \varepsilon_{\max} - \varepsilon$$
$$\text{With } \varepsilon_{\max} = 2.99\mu m \quad (2)$$

The results of calculations for estimating the reliability indices and failure probabilities are shown in table 4.

Table 4. Reliability results

	MCS(1000 samples)	Form	Sorm
CTE Ferrite	-	1.2232e -5	1.2232e -5
β	-	1.21	1.21
P_f	0.12	0.1131	0.1131
Time	23h	16min	23min

6. Conclusion

In this article we analyzed the thermal and thermomechanical behavior of a power inductor for amperage of *50A*. Firstly heat generation is computed in the coil and then it is completely simulated in ANSYS. Next, experimental tests were carried out using thermal measurements via thermocouples and mechanical measurements by an electronic speckle pattern interferometry, to measure deformation on the top and the side of the ferrite core.
On the basis of numerical and experimental results, we conclude that finite element model developed in ANSYS correctly describes the thermal and thermomechanical behavior of the power inductor.
The approach adopted allowed to establish a good base to perform a reliability study, a view to calculating the probability of failure with the methods of Monte Carlo and Form / Sorm.

Acknowledgment
This project has been funded with the support from the European Union with the European Regional Development Fund (ERDF) and from the Regional Council of Normandie. Project : CRIOS

References

[1] Goldman A 2006 *Modern Ferrite Technology* (Springer).
[2] Lemaire M 2005 *Structural reliability* (Lavoisier).
[3] Borza D, Gautrelet C 2015 Correlated full-field and pointwise temporally resolved measurements of thermomechanical stress inside an operating power transistor *Opt. Las. Eng.* **64** 79-88.
[4] Stetson K A, Brohinsky W R 1987 Electro-optics holography system for vibration analysis and nondestructive testing *opt. Eng.* **26** 1234-39.
[5] Stetson K A, 'et al' 1986 An electro-optic holography system with real-time arithmetic processing *J. Nond. Eval.* **8** 69-76.
[6] Nistea I 2010 *Développement des techniques optiques et acoustiquesde mesure 215 de champs orientées vers la vibroacoustique* Ph.D. Thesis INSA de ROUEN.
[7] Costantini M 1998 A novel phase unwrapping method based on network programming *IEEE Trans. Dev. Mat. Rel.* **36** 813-21.
[8] Luong B 2009 Costantini phase unwrapping, Tech. rep. Matlab Central - file.

IOP Conf. Series: Materials Science and Engineering **205** (2017) 012015 doi:10.1088/1757-899X/205/1/012015

Mechanical Property Analysis on Sandwich Structured Hybrid Composite Made from Natural Fibre, Glass Fibre and Ceramic Fibre Wool Reinforced with Epoxy Resin

K R Bharat[1,*], S Abhishek[1] and K Palanikumar[1]

[1]Department of Mechanical Engineering, Sri Sairam Institute of technology, Sai Leo Nagar, West Tambaram, Chennai, India
[*]E-mail: bharatrajasekaran@gmail.com

Abstract. Natural fibre composites find wide range of applications and usage in the automobile and manufacturing industries. They find lack in desired properties, which are required for present applications. In current scenario, many developments in composite materials involve the synthesis of Hybrid composite materials to overcome some of the lacking properties. In this present investigation, two sandwich structured hybrid composite materials have been made by reinforcing Aloe Vera-Ceramic Fibre Wool-Glass fibre with Epoxy resin matrix and Sisal fibre-Ceramic Fibre Wool-Glass fibre with Epoxy resin matrix and its mechanical properties such as Tensile, Flexural and Impact are tested and analyzed. The test results from the two samples are compared and the results show that sisal fibre reinforced hybrid composite has better mechanical properties than aloe vera reinforced hybrid composite.

1. Introduction

Stacking of several layers of different composite materials with many layers of natural and man-made fibre is a base for the development of hybrid composites. The main intention of using hybrid composite is to overcome the lacking properties of the fibres over the other in a single matrix material. In economical way, the unique features of hybrid composite have paved a way to meet the design requirements in a more subtle way, than the conventional materials. Natural fibres from plants [1] are reinforced with thermosetting plastics and they find their application in various industries, automobile sectors and so on. The surface properties of these natural fibres are enhanced by subjecting them to chemical treatments [2]. The fine fibres of glass are intermingled in mat form and they are reinforced with polymer matrix to get light weight robust material for industrial, automobile and aerospace application.

The ceramic fibre wool, a mineral fibre made by man and the main constituents present in it are alumina and silica. It has some phenomenal energy saving property, that dissipates heat from the furnace, so it has low thermal conductivity. The glass fibre and various natural fibre reinforced hybrid polymer composites are studied by cicala et al [3]. They have observed the properties and performance of these hybrid composites for the application of curved pipes. Idicula et al[4] have studied the thermo physical properties of natural fibre reinforced polyester composites and they found that, the heat transport ability of these composites are improved due to the reinforcement of polyester composite made with natural fibre along with glass fibre. Ye han et al[5] have studied the friction performance of ceramic fibre wool for automotive brake lining application and they endow that, wear rate of the material is increased and the heat fading resistance property of the material is improved because the aluminum and silicon contents present in the wool have increased to 5 % weight. In this study, two different composition of hybrid composite specimens are prepared and subjected to mechanical property testing such as tensile, flexural and impact test. The test results and graphs of these two hybrid composite specimens are compared and described in detail. The results prevails that, one of the hybrid composite specimens has better mechanical property than the other.

2. Experimental

2.1. Materials

Aloe vera (Aloe Barbadensis), Sisal (Agave sisalana), Ceramic fibre wool and E-Glass fibre are the core materials based on which, two different hybrid composites are prepared and tested. Aloe vera and Sisal fibre mat from Jute Weaver Association, Anakaputhur, Ceramic fibre wool from B.M. Insulations Pvt. Ltd, Gerugambakkam and E-Glass fibre from Sakthi Fibres, Saidapet are procured and suitable binders for these fibres are chosen which include Epoxy Resin of grade LY 556 and hardener of grade HY 906, which are procured from Sakthi fibres . Fabrication and Mechanical Testing of this specimen are carried out at L.M.P R&D, Laboratory, Pallipalayam, Erode, Tamil Nadu, India.

2.2. Hybrid Composite preparation

Two hybrid composite specimen are sandwiched to form the resultant hybrid composite of 300 x 300 mm dimension with 5.5 mm thickness. Sandwich structured hybrid composite specimen incorporates an innermost layer of Ceramic fibre wool covered by an intermediate double layered E-glass fibre, which again is enclosed by a double layered woven Aloe vera in the former and Sisal fibre in the later. Epoxy resin of LY 556 grade is mixed with HY 906 grade hardener at the proportion of 10:1. On each layers of the hybrid composite, a required amount of epoxy resin is applied on both fibre surfaces and then reinforcement is done in compression moulding machine. For curing of specimen, 100°C temperature for duration of 30 minutes is maintained within the machine. After the curing process, the specimen is taken out from the compression moulding machine.

2.3. Specimen preparation

The cured specimen taken from the compression moulding machine is properly barbered to carry out mechanical property testing such as Tensile, Flexural and Impact as per ASTM standard. The images of the barbered specimen before and after testing are shown in Figure 1.

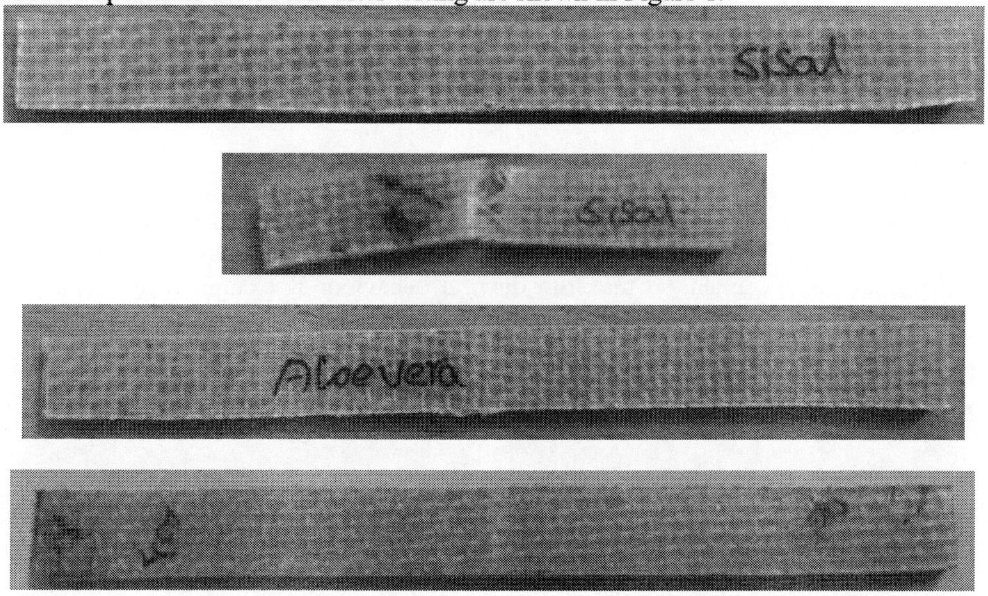

Fig. 1 Specimen used before and after the testing

2.4. Mechanical properties

2.4.1. Tensile testing

The Universal Testing Machine of model number KIC-2-1000-C is used to carry out tensile test on the barbered specimen as per ASTM-D3039 standard [6]. At a speed of 2 mm/min, six strip shaped samples of (dimension 250 x 25 x 5.5 mm) specimen are tested and then, the values are taken and the graph (stress-strain and load-displacement) generated by the machine is shown in the figures 2 and 3.

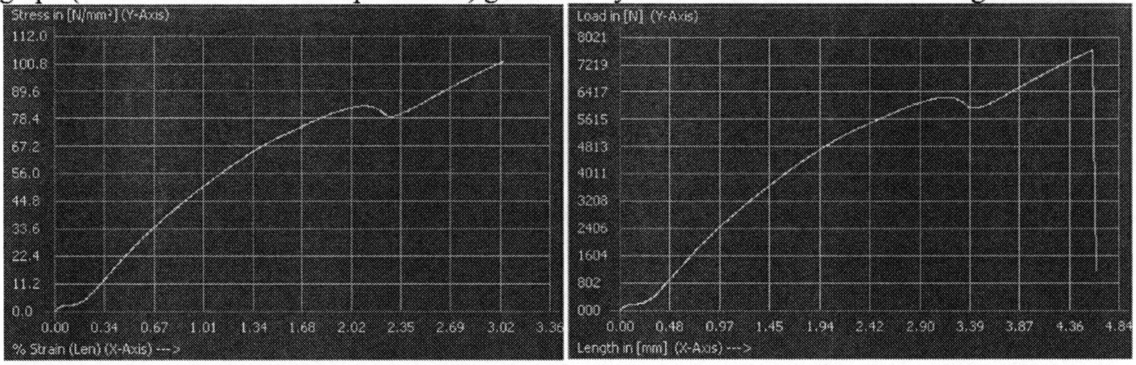

Figure 2. Stress-Strain curve and Load-Displacement curve are generated by UTM for Tensile test of Sisal fibre + E-Glass fibre + Ceramic fibre wool Hybrid composite.

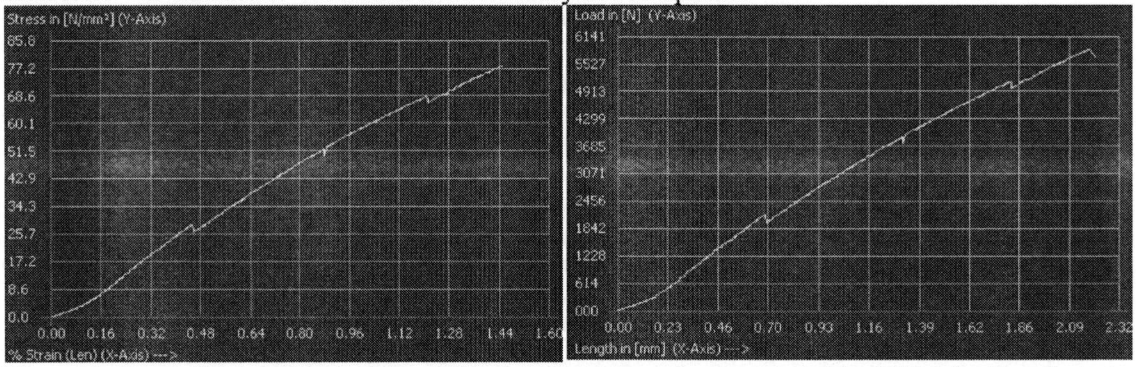

Figure 3. Stress-Strain curve and Load-Displacement curve are generated by UTM for Tensile test of Aloe vera fibre + E-Glass fibre + Ceramic fibre wool Hybrid composite.

2.4.2. Flexural testing

As per the ASTM-D790 standard [7], six samples of (dimension 125 x 13 x 5.5 mm) of two different composition specimen are subjected to three point testing. The compressive and tensile forces are experienced by the upper and lower surface of the specimen and hence, the shear force is induced in the central part of the specimen, so that, the flexural behaviour is examined up to its yield point. During testing, the stress-strain graph generated by the machine is shown in figures 4 and 5.

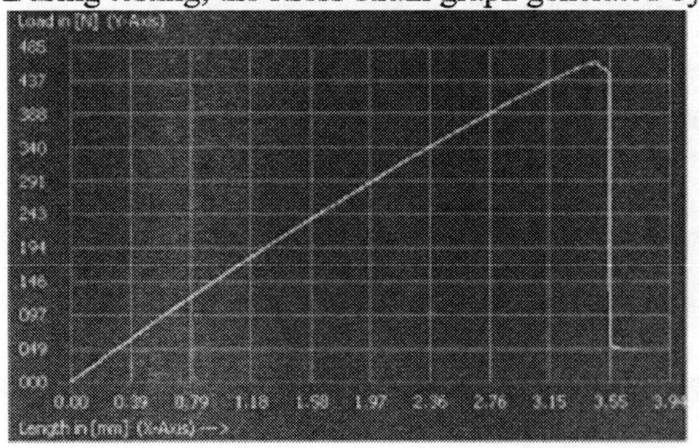

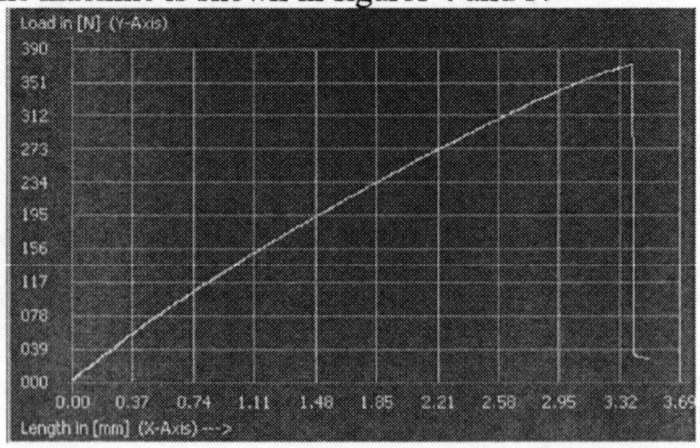

Figure 4. Stress-Strain curve is generated by UTM for Flexural test of Sisal fibre + E-Glass fibre + Ceramic fibre wool Hybrid composite.

Figure 5. Stress-Strain curve is generated by UTM for Flexural test of Aloevera fibre + E-Glass fibre + Ceramic fibre wool Hybrid composite.

2.4.3. Impact testing

The hybrid composite specimen samples of two different compositions mentioned above are subjected to heavy impact load to measure the impact force of the specimen against rupture. The testing was carried out as per the ASTM-D256 standard [8] for three specimens of (dimensions 65 x 13 x 5.5 mm) two different compositions to get an average value and the specimen's image is shown in figures 1.

3. Result and Discussion

Nowadays, hybrid composite materials are used in various field, because their mechanical properties have surpassed the conventional materials. Mechanical testing's such as Tensile, Flexural and Impact test on the two different hybrid composite are performed and their mean average values are correlated with each other, which are described in the Table 1, where Sample 1 represents Sisal fibre + E-Glass fibre + Ceramic fibre wool Hybrid composite and Sample 2 represents Aloe vera fibre + E-Glass fibre + Ceramic fibre wool Hybrid composite.

Table 1. Mechanical properties of tested Hybrid composites.

Composite sample	Tensile properties			Flexural properties		Impact strength (J)
	Peak load (KN)	Ultimate Tensile Strength (MPa)	% Elongation	Flexural load (KN)	Flexural Strength (MPa)	
Sample 1	7.639	101.857	3.067	0.461	110.951	1.90
Sample 2	5.848	77.98	1.467	0.371	89.35	1.75

3.1. Tensile properties

In Universal Testing Machine, six samples from two different compositions of hybrid composite specimen are tested until rupture and the ultimate tensile strength of the specimens is determined. While testing the specimen, at peak load the rupture takes place, along with percentage elongation is determined and the graph generated by the machine for stress-strain and load-displacement is plotted and they are shown in the figures 2 and 3 respectively. From the figure 2, load-displacement curve of Sample 1 has the high value of ultimate tensile strength of 101.857 MPa than the Sample 2 of 77.98 MPa. Also, the peak load of Sample 1 value 7.639 KN is higher than the Sample 2 value 5.848 KN. Similarly, the stress-strain curve showed in the Figure 2 for the above two samples shows the same result as that of the load-displacement curve in which the percentage elongation of Sample 1 value 3.067 is higher than Sample 2 value 1.067.

3.2. Flexural properties

The flexural strength for Sample 1, with the value of 101.961 MPa is higher than Sample 2 value, 89.35 MPa and the flexural load is higher for Sample1 value, 0.461 KN than the Sample 2 value, 0.371 KN and this is perceived from the Table 1. The stress-strain curves for the above two samples reveals the same as shown in Figures 4 and 5. The measure for the flexural strength is stiffness. The stiffness of the Sample 1 is increased to a little extend than Sample 2, because of the proper binding force between the fibres, wool and matrix of Sample 1 than Sample 2.

3.3. Impact properties

The capability of impact for the above mentioned specimens is carried out in Izod impact machine and the energy loss of the specimens is found out through the reading given by the machine. The response of impact leads to the formation and development of crack and breakage of resin, fibre and wool interface. The readings from the machine are presented in the Table 1 and it clearly shows that the impact strength for Sample 1 value, 1.90 J is little higher than Sample 2 value, 1.75 J. This prevails that, the resistivity of Sample1 is higher than Sample 2.

4. Conclusion

In this present study, one type of specimen is made of Sisal fibre, E-glass fibre, Ceramic fibre wool and epoxy resin matrix and another one is made of Aloe vera fibre, E-glass fibre, Ceramic fibre wool and epoxy resin matrix. These are fabricated and subjected to mechanical property test such as tensile, flexural and impact test. The test results of these two specimens are compared and then, the conclusions are drawn:

- The tensile test prevails that sample 1has higher ultimate tensile strength value, 101.857 MPa and higher peak load, 7.639 KN.
- The sample1 has the capability of higher flexural strength value, 110.951 MPa and higher flexural load, 0.461 KN than sample 2.
- Impact strength is high for sample 1 with the value of 1.90 J, while, sample 2 has lower impact strength of 1.75 J.

5. References

[1] Saravana Bavan D, Mohan Kumar G C. *Potential use of natural fibre composite materials in india.* Journal of Reinforced Plastics and Composites 2010 **29:** 3600.

[2] Bongarde U S, Shinde V D. *Review on natural fibre reinforcement polymer composites.* International Journal of Engineering Science and Innovative Technology 2014. Vol. **3**, issue 2.

[3] Cicala G, Cristaldi G, Recca G, Ziegmann G, El-Sabbagh A, Dickert M, et al. *Properties and performance of various hybrid glass / natural fibre composites for curved pipes.* Mater Des 2009; **30**:2538-42.

[4] Idicula Maries, Boudenne Abderrahim, Umadevi L, Ibos Laurent, Candau Yvess, Thomas Sabu. *Thermophysical properties of natural fibre reinforced polyester composites.* Compos Sci Technol 2006;**66**:2719-25.

[5] Ye Han, Xiaofeng Tian, Yansheng Yin. *Effect of ceramic fibre on the friction performance of automotive brake lining materials.* Tribology Transactions 2008; vol-**51**, issue-6.

[6] ASTM D3039 Standard test method for testing tensile properties of plastics.

[7] ASTM D790-07 Standard tests method for testing flexural properties of unreinforced and reinforced plastics and electrical insulating material.

[8] ASTM D256-06a Standard test method for determining izod pendulum impact resistance of plastics.

[9] Ramesh M, Palanikumar K, Hemachandra Reddy K. *Mechanical property evaluation of sisal jute-glass fibre reinforced polyester composites.* Composite: part B **48** (2013).

[10] Shadrach Jeya Sekaran A, Palanikumar K, Pitchandi K. *Evaluation of mechanical properties of woven aloe vera and sisal fibre hybrid reinforced epoxy composites.* Bull. Mater. Sci. vol **38**, issue 5.

IOP Conf. Series: Materials Science and Engineering **205** (2017) 012016 doi:10.1088/1757-899X/205/1/012016

Composite Shell Strain Detection for SRM Based on Optical Fiber Sensors

Zhang Lei, Chang Xin-Long, Zhang You-hong, Chen Xiang-dong

Xi'an Hi-Tech Institute, Xi'an, Shanxi 710025, China
lwzy_zl@163.com

Abstract. As a new passive sensor, fiber Bragg grating (FBG) sensors have provided a new idea for the SRM shell damage detection, which is to integrate the FBG sensor network in the material interior or to the surface to monitor the shell structure. However, it is difficult to embed the FBG sensor in filament wound composite material structure for the reason of large tension and high temperature in process of manufacture. Therefore we propose a new method that embed FBG sensor network between the composite shell surface and the thermal protective coating. The calibration of sensor is presented by tensile test and the strain transfer coefficient is gotten. It is certified by the hydrostatic test that the FBG sensors could precisely describe the strain variation and distribution of the composite shell and effectively improve the survival rate by embedding the FBG sensors between the composite shell surface and the thermal protective coating.

1. Introduction

As an important part of missile weapon system, the health status of Solid Rocket Motor (SRM) composite shell structures is of great concern to the system. At present, the shell of solid missile in service around the world is wide application of filament wound composite material structure which has lots of advantages such as light quality, lower cost and higher than strength and stiffness, corrosion resistance and good stability, etc. However, composite materials have heterogeneity, anisotropy and process discreteness, easily appear internal defects such as pores, layered in the process manufacturing. Due to the collision in the process of transportation, storage, environmental erosion, and aging factors, the shell may generate much defect such as fracture, layering, and even delamination. In recent years, optical fiber sensors [1-7] have shown greater benefits in the field of static and dynamic strain monitoring with respect to more traditional sensors, such as immunity to electromagnetic fields, durability, resistance to corrosion and multiplexing capability. Furthermore, their small sizes and compatibility with common composites materials make them easily embeddable inside the structure of solid rocket motors (SRM) without inducing significant weakening of the structure. However， the main research on damage detection all concentrated on the structure of laminated plates which is easy to embedded the FBG and the survival rate is high. Due to the filament wound tension and high temperature of curing during the course of manufacture ， the embedded FBG sensors is easily overloaded and damaged[8]. Also, the Bragg spectrum may be distorted depending on the intensity and spatial distribution of the residual strain field and finally the strain transfer from the host material to the fiber core is different in the case of an embedded sensor rather than an adhesively bonded sensor. Another problem is that the location of the FBG sensors is difficult to determine in processing of fiber winding.

In this paper, we propose a new method that FBG sensor network was embedded between the composite shell surface and the thermal protective coating. The strain detection system to detect the

IOP Conf. Series: Materials Science and Engineering **205** (2017) 012016 doi:10.1088/1757-899X/205/1/012016

shell strain distribution under pressure and the feasibility of composite shell strain detection based on FBG sensors was verified by hydrostatic test of small scale SRM shell . This way of embedment effectively improve the survival rate of the FBG sensor.

2. Basic of FBG Operation

A fiber Bragg grating is a periodic modulation of the effective refractive index n_{eff} in the optical fiber core. Light travelling at the Bragg wavelength, λ_B, which depends on the grating features, is reflected back by the grating itself and as a result it is missing in the transmission spectrum. The reflected spectrum is centered at the Bragg wavelength and depends on the mentioned effective index of refraction (n_{eff}) and on the Bragg period (Λ) of the grating, according to the following equation:

$$\lambda_B = 2n_{eff}\Lambda \qquad (1)$$

which is the Bragg equation. By inscribing an array of FBG sensors with different grating periods in the same fiber, it is possible to monitor different locations on a structure. When a local deformation is present, the grating's period varies and the reflected wavelength varies accordingly, allowing the detection of the local strain via the Bragg equation (1). The strain and the wavelength variations can be put in a one to-one relationship only in the case of unidirectional and uniform deformation, provided that the effect of the temperature on the Bragg wavelength is suitably compensated for.

3. FBG Sensors Embedded and Calibration

After the shell curing and molding, the FBG sensor arrays were arranged by means of surface paste on the shell surface. Then the thermal protective coating was sprayed evenly on the surface. In this way the sensor arrays were embedded between the shell and coating. When FBG sensors embedded between missile composite shell and coating, the adhesive and coating all will influence the strain transfer ratio of the sensor. Because the adhesive thickness is also difficult to precise control and so it is difficult to calculate by the strain transfer function of the strain transfer ratio of the sensor. In order to build a practical application of optical fiber sensing test system, and finally realize in the application of solid rocket motor, it is very necessary to calibrate relationship between the shift of FBG center wavelength and the strain of composite shell before the actual application.

3.1. Calibration method

The calibration test was completed by axial tensile test. Axial tensile specimens were made of glass fiber composites according to Chinese standard of P.RC, GBT1447The length of Bragg grating is 1.5cm. Stick the FBG by adhesive on the specimen surface along the tensile direction . Then the coating was evenly sprayed on the surface of specimen. High precision resistance strain gages used to measure the axial real strain was pasted on the other side of the specimen surface and ensure the paste direction in strict accordance with fiber grating direction. The demodulation system of the FBG in test is SM130 produced by MOI. The calibration test as figure 1 shown.

3.2. Test data processing and analysis

In the process of test, the high precision resistance strain gauge and fiber demodulation instrument were used to track measurement and record of strain gauge and data at the same time. Due to the large strain taken place on tensile specimen, the FBG demodulation instrument hasn't demodulated the abnormal wave shape when the tensile strain reach 7932 $\mu\varepsilon$. The test finished by now. The strain gauge and fiber sampling frequency demodulation devices are set to 1 Hz. To calibration of sensor, need to get a grating center wavelength variation - strain curve. As the figure 2 shown, strain-time and wavelength-time data were depicted in a coordinate system.

To calibration of sensor, we need to get a grating center wavelength variation-strain curve. A new coordinate system with the strain gauge measured strain values as X axes and the sensor demodulation instrument center wavelength variation as Y axes was established. Then the center wavelength variation of the sensor and strain curve was obtained, as shown in figure 3. The real strain measured by FBG and gauge measurement were compared, as shown in figure 4.

IOP Conf. Series: Materials Science and Engineering **205** (2017) 012016 doi:10.1088/1757-899X/205/1/012016

Figure 1. Calibration test

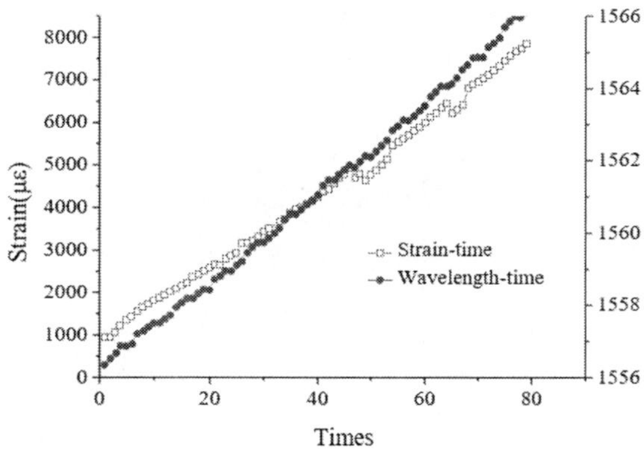

Figure 2. strain-time and wavelength-time data in tensile test

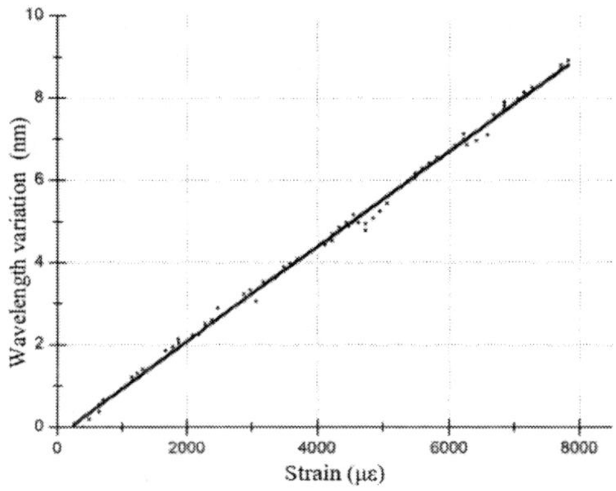

Figure 3. Strain- wavelength variation curve

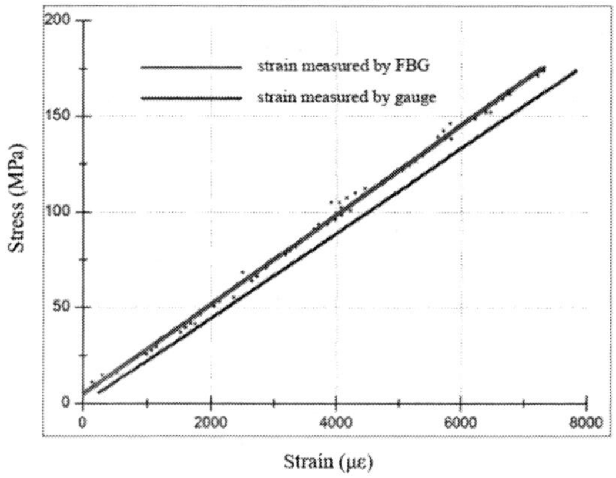

Figure 4. Comparison strain measured by FBG and resistance strain gauge

Linear regression analysis for date fitting was made by use of the mathematical tools Origin software, the strain-wavelength linear regression equation is obtained as follows:

$$y=0.001151x-0.22748 \quad (2)$$

The degree of fitting:

$$R^2=0.99746 \quad (3)$$

The average sensitivity:

$$S=\Delta\lambda/\Delta\varepsilon=1.151 pm/\mu m \quad (4)$$

Where $\Delta\lambda$ is FBG center wavelength variation, $\Delta\varepsilon$ is the strain variation. It can be conclude that the FBG sensor keep a good linearity and sensitivity.

4. Hydrostatic Test of Small Scale SRM Shell

In order to test the feasibility of the sensor network with the FBG embedded between composite and coating, a glass fiber filament wound scale SRM model embedded sensors was used to hydrostatic test. The length and radius of cylindrical section of the shell are r= 75 mm, and l = 160 mm, respectively. The staking sequences of 9 laminae (from inner to outer) is $[(\pm 28°)/(90°)_2]_3$. As the figure 5 shown, small scale SRM shell coated by thermal protective coating .

Figure 5. Small scale SRM shell coated by thermal protective coating

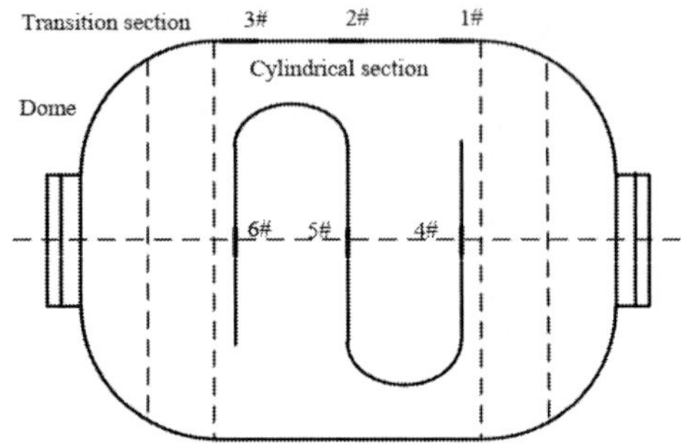

Figure 6. Arrangement of FBG sensors

4.1. Sensors arrangement and hydrostatic test

The sensors arrangement are as follows: In the cylindrical section along the axial deploy a three series grating, along the hoop direction deploy a three series grating, then two quasi distributed sensor network has formed on the shell surface. As the figure 6 shown ,the FBG sensor 1 #, 2 #, 3 # are along the axial direction, their wavelength is 1550.155 nm ,1555.102 nm and 1560.126 nm respectively, distance between each grating is 5 cm; 4 #, 5 #, 6 # are along the hoop direction, the wavelength of 1530.872 nm , 1535.728 nm and 1539.760 nm, distance between each grating is 5 cm also. After arrangement the sensor, the coating was evenly sprayed on the surface of the cylindrical section of the shell. After the coating fully cured, the FBG sensors were tested by using of demodulation instrument. We can see that the survival rate of FBG sensor effectively improve and reach 100%. In order to verify the test result, six high precision resistance strain gages were pasted near the location of the FBG on the surface of the shell, the test strain direction of strain gages are consistent with the FBG sensor.

The water pressure slowly increase from 0 to 5Mpa, keep the pressure for 10s . Then release the pressure gradually until extinction, record the process of centre wavelength and grating strain gauge measured strain values.

4.2. Test data processing and analysis

Due to the shell was knocked down accidentally in the process of hydrostatic test, an axial arrangement with FBG sensor and strain gauge were failure, at the end of the test only five sets of data were collected. Recording the test data, we can get the relationship between centre wavelength and time. According to the strain-wavelength linear regression equation (2), the relationship between strain measured by FBG sensors and time can be gotten. Figure 7 and Figure 8 respectively show strain measured by FBG and resistance strain gauge in the process hydrostatic test. In according to the comparison of the two figures, we can see that the trend of two sets of curves are basically identical. Experimental results demonstrate the good agreement between the concentration strain filed provided by strain gauges and that provided by the FBG sensors network. However, the FBG sensor can be more precisely the change of strain on the shell than the resistance strain gauge. In the hydraulic testing, the hoop strain was bigger than axial strain in the cylindrical section. In the middle of cylindrical section the hoop strain is biggest than the strain of other section pasted sensors.

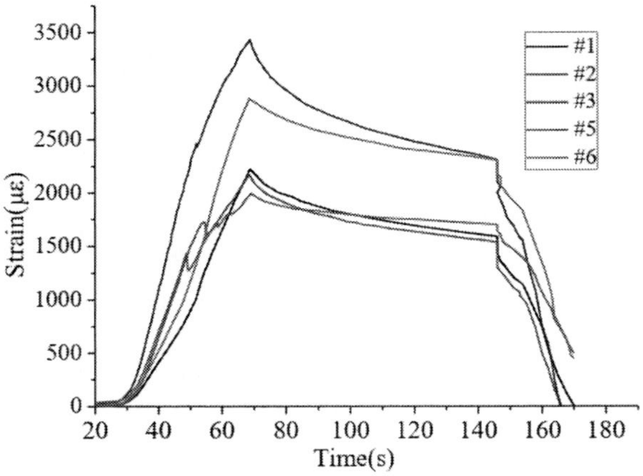

Figure 7. Strain measured by resistance strain gauge

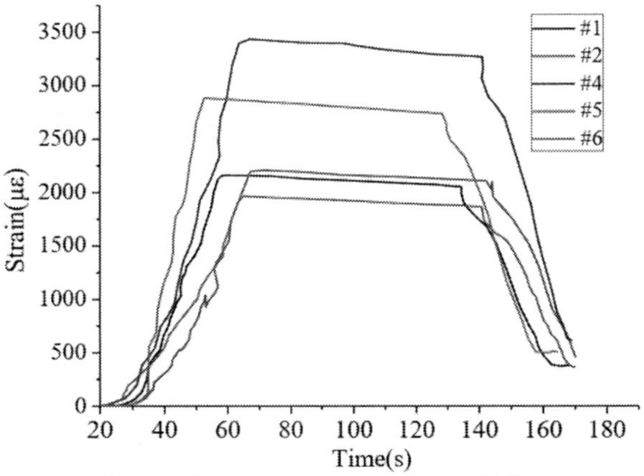

Figure 8. Strain measured by FBG

5 Conclusion

In order to realize the strain monitoring of FBG sensors in SRM, a method of embedding FBG sensor between composite shell and the thermal protective coating is proposed. This method can effetely improve the survival of FBG sensor than embedding directly into the composite fiber of shell and achieves 100% survival in the Hydrostatic test. It is certified by the hydrostatic test that the FBG sensors could precisely describe the strain variation and distribution of the composite shell and effectively improve the survival rate by embedding the FBG sensors between the composite shell surface and the thermal protective coating. As a first step research for the development of health monitoring system for SRM, the capability of embedded FBG sensors network system in composite structures has been demonstrated in these experiments.

References

[1] W. R. Habel, Reliable Use of Fiber Optic Sensors. In:Encyclopedia of Structural Health Monitoring, Eds: Boller C. etal., John Wiley & Sons (2009).

[2] Costiner S, Winston H A, Gurvich M R, Ghoshal A, Welsh G S, Butler S L, Urban M R, Bordick N . Optimal Sensor Fusion for Structural Health Monitoring of Aircraft Composite Components[R] . United Technologies Research Center,2011:1-9.

[3] Degrieck J, De Waele W, Verleysen P. Monitoring of fibre reinforced composites with embedded optical fibre Bragg sensors, with application to filament wound pressure vessels. NDT&E International 2001;34(4):289–96.

[4] Grant J, Kaul R, Taylor S, Myers G, Wilkerson C, Jackson K, et al. Distributed sensing of carbon-epoxy composites and filament wound pressure vessels using fiber-Bragg gratings. Proc SPIE 2002;4935:32–40.

[5] Kunzler M, Udd E, Kreger S, Johnson M, Henrie V. Damage evaluation and analysis of composite pressure vessels using fiber Bragg gratings to determine structural health. Proc SPIE 2005;5758:168–76.

[6] Ortyl Nicholas E. Damage evaluation and analysis of composite pressure vessels using fiber Bragg gratings to determine structural health. Proc SPIE 2005;6004.

[7] Kang DH, Kim CU, Kim CG. The embedment of fiber Bragg grating sensors into filament wound pressure tanks considering multiplexing. NDT&E Int 2006;39(2):109–16.

[8] Chang, Xinlong; He, Xiangyong; Hu, Jianghua; Li, Jinjun. Experimental research on embedded fiber Bragg grating sensors network for solid rocket motors health monitor[C]. Proceedings - The 1st International Conference on Intelligent Networks and Intelligent Systems, ICINIS 2008.

Intermetallic Growth Retardation with the Addition of Graphene in the Sn-3.5Ag Lead-free Solder

R Mayappan[1], A Salleh[2]

[1] Faculty of Applied Science, Universiti Teknologi MARA, Perlis Campus, 02600 Arau, Perlis, Malaysia
[2] Faculty of Applied Science, Universiti Teknologi MARA, 40450 Shah Alam, Selangor, Malaysia
amirahsalleh90@gmail.com

Abstract. In this study, the effect of 0.07 wt.% graphene nanosheets (GNSs) on the thermal, microstructure and intermetallic compounds (IMC) growth kinetics were investigated. Experimental results showed that addition of GNSs to the Sn-3.5Ag solder alloy slightly increased the melting temperature. Scanning electron microscope (SEM) micrograph revealed that with the addition of GNSs has transformed the morphologies of the interfacial IMC from scalloped to more planar-shaped after isothermal aging for 500 hours. Growth kinetics calculations show that the k value was reduced from 9.08 x 10^{-14} cm^2/s to 5.73 x 10^{-14} cm^2/s with the addition of graphene. Thereby, it indicates that the addition of GNSs suppressed the growth of the IMC layers hence retarded the IMC's growth.

1. Introduction

In recent times, many lead-free solders have been developed in the electronic packaging industry as a replacements for the conventional Sn-Pb solder even though the Sn-Pb solder has been widely used due to their lower melting temperatures and good wetting behavior. Due to the harmful effect to human health and legislative pressures, the electronic industry is refocusing toward implementation of lead-free solders. Among them, the Sn-Ag alloy systems with or without the additions of other alloying elements is considered as one of the best promising candidate because of its mechanical and electrical properties [1-2].

Nevertheless, one of the main problem related to the use of Sn-3.5Ag is an excessive reaction between the solder and substrate which results to large brittle formation of intermetallic compounds (IMC) [3]. IMC is a layer that generated by chemical reaction between solder alloy and substrate during the soldering process which can grow via diffusion in the solid state. For the Sn-Ag/Cu couple, the interfacial IMC layer consists of Cu_6Sn_5 and Cu_3Sn binary compounds and these layer were generated during the soldering process and grew in the high temperature environment [4]. Many studies have been done to further improve the solder properties in order to reduce the IMC growth by adding extra element such as In, Bi, and Zn into the solder alloys [5-8]. Recently, nano-composite solders have been developed due to rapidly changes in technologies and moving towards miniaturized of consumer electronic products including the size of electronic components. The reinforcements of nano-particles to the solder alloys could result in self-organized dispersive systems, which contribute to the heterogeneous nucleation, thus causing the stress in the solder joints to be dispensing uniformly [9-10]. Among all the micrometer and nanometer sized particles as a new reinforcement, graphene has attracted the attention of researchers due to its excellent mechanical, electrical and thermal properties. Few studies [11-13] reported that addition GNSs into the pure solder has showed improvement of mechanical properties. Thus, this present work investigated the effects addition of 0.07 wt.% graphene nanosheets (GNSs) into Sn-3.5Ag solder on thermal and microstructure formation of IMC. In view of this information, the isothermal aging were done to study the reactions between the solders and Cu substrate to understand the formation and growth mechanism of IMC.

2. Methodology.

2.1. DSC measurement for melting point

Melting characteristics of the Sn-3.5Ag and Sn-3.5Ag-0.07GNSs solders were investigated using differential scanning calorimeter (Jade DSC Perkin-Elmer). The solder samples were heated to 250 °C at heating rate of 10 °C/min. The entire scanning was carried out under inert nitrogen atmosphere.

2.2. Interfacial study of the IMC

Composite solder was prepared by powder metallurgy route method by mixing 0.07 wt.% of graphene nanosheets (GNSs) into Sn-3.5 wt.%Ag solder. The mixtures were blended mechanically for 2 hours for Sn-3.5Ag and 10 hours for Sn-3.5Ag-0.07GNSs with 200 rpm speed to achieve uniform distribution. A similar mixing method was used by other researchers [14]. The range diameter of the GNSs used is 0.5–2.0μm and 25 nm of thickness. Then the mixtures were compacted into disc shape with a diameter of 12 mm at 14 MPa and sintered at 150 °C for 2 hours in the argon gas atmosphere. For the study of interfacial reactions, the Sn-3.5Ag and Sn-3.5Ag-0.07GNSs solders were placed on the Cu substrate and molten at 250 °C for one minute. The samples were isothermally aged at 150 °C for 100, 200, 300, 400 and 500 hours.

The solder joint was mounted in epoxy, cross-sectioned, polished with silicon carbide (SiC) papers until 2000 grit and final polishing was done with 1.0 and 0.3 μm silica powders, and etched for 1 seconds to reveal the intermetallic microstructure. The Cu/solder interface was then analysed under Scanning Electron Microscope (SEM) and the average thickness of the intermetallic was calculated by using imageJ software.

3. Results and Discussions.

3.1. Thermal behavior

Differential scanning calorimetry (DSC) analysis was done to investigate the fundamental thermal effect during heating process of these two solders. Figure 1 illustrate the DSC curves of Sn-3.5Ag and Sn-3.5Ag-0.07GNSs through heating process. During the endothermal process, only one eutectic peak was observed for Sn-3.5Ag solder, and the onset and peak temperatures appeared at 224.83 °C and 228.92 °C, respectively. With the addition of GNSs nanoparticles, the onset and peak temperatures slightly changed towards the right. The onset and peak temperatures of the Sn-3.5Ag-0.07GNSs increased to 225.03 °C and 230.20 °C, respectively. These results show that the addition of GNSs has a little effect on the T_{onset} and T_{peak} which is <1.3 °C. The reason could be that the reinforcing nano-sized of GNSs particles into Sn-3.5Ag solder changed the surface instability and variation in physical properties of grain boundary/interfacial characteristics. Such particles may serve as restrained sites for the solidification process of IMCs as well as β-Sn grains. The literature [15] shows that the addition of little amount of graphene oxide nanosheets has slightly increased the melting point of Sn-5Sb-1Ag nanocomposite solder. So, its influence on the melting point of composite solder can be ignored because of its low GNSs content as reported by some researchers [12].

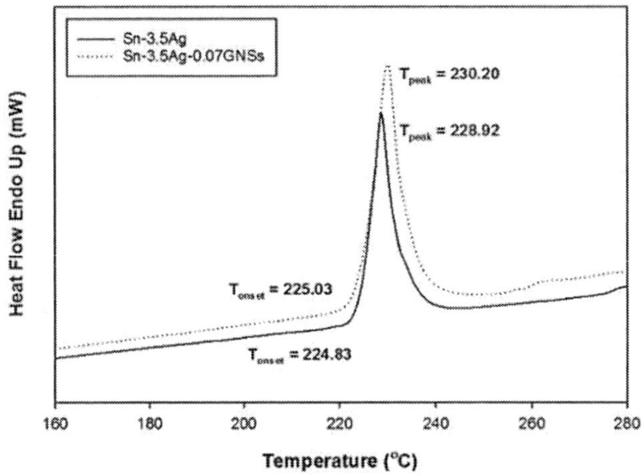

Figure 1. DSC result during heating for Sn-3.5Ag and Sn-3.5Ag-0.07GNSs solders.

3.2. Morphology of the Intermetallic.

The SEM micrographs of Sn-3.5Ag/Cu and Sn-3.5Ag-0.07GNSs/Cu are shown in figures 2 to 5. Figures 2a and 2b present the solder joint after reflowed for 1 minute at 250 °C, in which the interfacial Cu_6Sn_5 IMC layer is very thin (average about 2.1 ~ 2.4 μm), scallop-like shaped and appeared in light gray layer. Based on previous study [16], a single layer of Cu_6Sn_5 intermetallic compound layer was also observed at the solder/Cu interface during the soldering process of Sn-Pb and Sn-Zn solders.

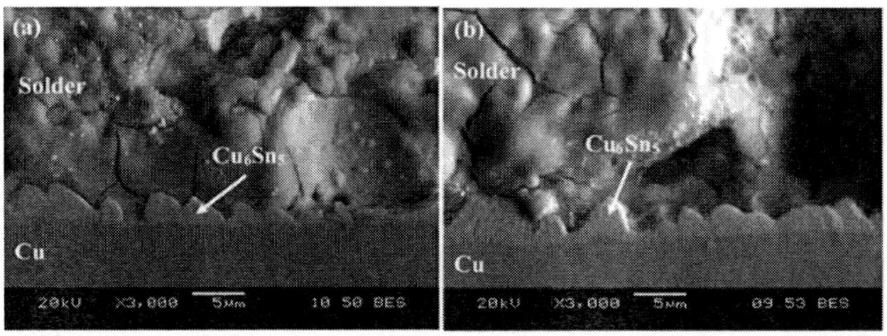

Figure 2. IMC formation for as-soldered for 1 min at 250 °C for (a) Sn-3.5Ag/ Cu and (b) Sn-3.5Ag-0.07GNSs/Cu solder joints.

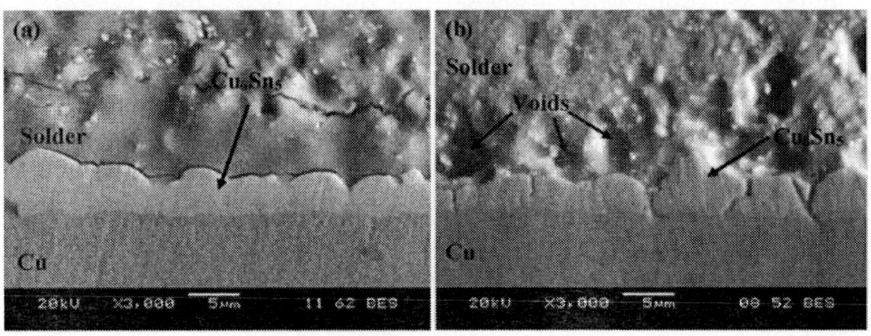

Figure 3. IMC formation for (a) Sn-3.5Ag/Cu and (b) Sn-3.5Ag-0.07GNSs/Cu solder joints after aged for 100 hours.

The IMC layers become thicker (3.5 – 6.5 µm) and turn to more planar-liked shape when aging time increasing (figures 3-5). When samples were aged for 300 hours, the Cu_3Sn layer was observed at the interface between Cu_6Sn_5 intermetallic compound and Cu substrate for the Sn-3.5Ag-0.07GNSs solder joint (figure 4b). Since the Cu_6Sn_5 intermetallic has lower activation energy compared to Cu_3Sn, so it was the first phase that formed at the substrate/solder interface while Cu_3Sn layer required more time to generate. As the aging time prolonged until 500 hours (figure 5) the Cu_3Sn layer was appeared for both Sn-3.5Ag and Sn-3.5Ag-0.07GNSs solder joints. The solder joints with the addition of graphene showed decreasing in intermetallic thickness formation at the interface compared to the Sn-3.5Ag solder joints. Some voids were also appeared in the solder interfacial for Sn-3.5Ag-0.07GNSs/Cu solder joints which may be due to the present of graphene nanosheets in the solder matrix.

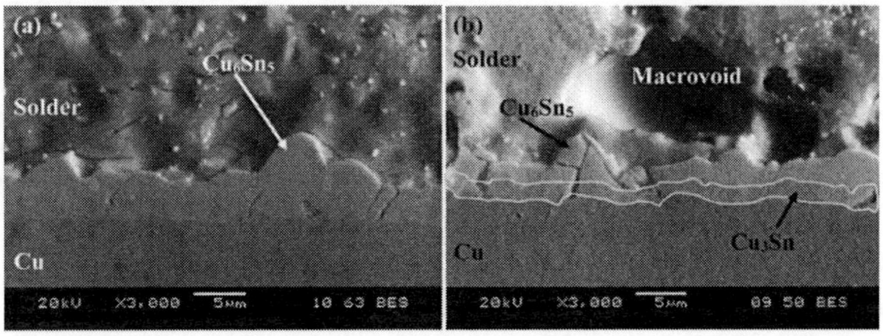

Figure 4. IMC formation for (a) Sn-3.5Ag/Cu and (b) Sn-3.5Ag-0.07GNSs/Cu solder joints after aged for 300 hours.

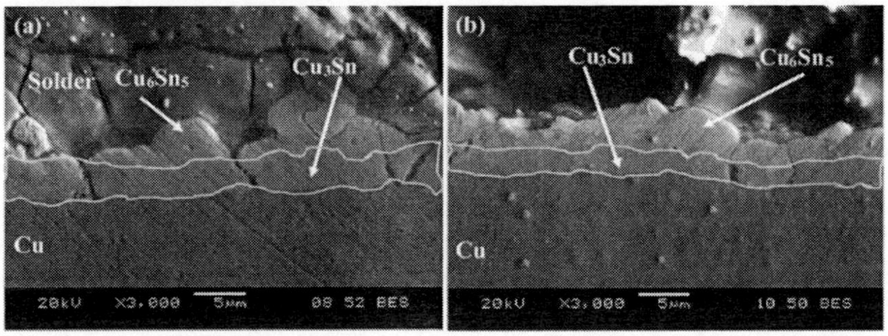

Figure 5. IMC formation for (a) Sn-3.5Ag/Cu and (b) Sn-3.5Ag-0.07GNSs/Cu solder joints after aged for 500 hours.

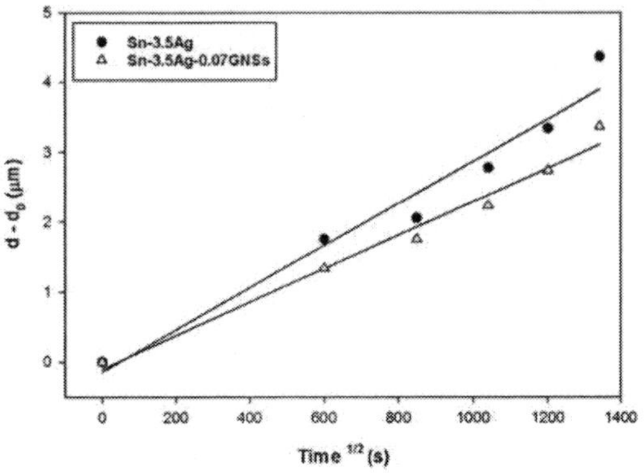

Figure 6. The d – d_0 of the IMC layer thickness with respect to different isothermal aging time.

The average values of thickness during aging, d, was plotted against the square root of aging time, t, as shown in figure 6. The expression for diffusion controlled mechanism is given by:

$$d - d_0 = k^{1/2}\, t^{1/2} \quad (1)$$

where d_0 is the initial thickness of IMC layer and k is growth rate constant. Using this relation, the value of k was calculated to be higher for Sn-3.5Ag which is 9.08 x 10^{-14} cm^2/s compared to 5.73 x 10^{-14} cm^2/s for Sn-3.5Ag-0.07GNSs solder. This indicates that the reinforcement of GNSs into Sn-3.5Ag solder has supress the growth rate of IMC formation between the solder and the substrate.

4. Conclusion

The melting temperature and interfacial microstructure of Sn-3.5Ag and Sn-3.5Ag-0.07GNSs solders were investigated. Addition of 0.07 wt.% of GNSs into Sn-3.5Ag solder slightly increased the melting temperature. A scallop layer of IMC was generated at the interface during the soldering. The morphology of IMC changed from scallop-like to irregular shape IMC during isothermal aging till 500 hours for both solder joints. The Sn-3.5Ag-0.07GNSs solder joints in general exhibit lower growth rate as compared to Sn-3.5Ag solder joints. It indicated that GNSs can effectively retard the IMC's layer growth.

Acknowledgement

This study is financially supported by The Minister of Higher Education (MOHE) for the Grant 600-RMS/FRGS 5/3 (104/2015). Authors also thank Universiti Teknologi MARA and UniMAP for facilities.

References

[1] Hu X W, Xu T, Jiang X X, Li Y L, Liu Y and Min Z X 2016 *Appl. Phys.* **A 122** 278
[2] Yao Y, Wang Y X, Keer L M and Fine M E 2015 *Scripta Materialia* **95** 7-10
[3] Gain A K, Chan Y C and Yung W K C 2011 *Microeletron. Reliab.* **51** 975
[4] Bader S, Gust W and Hieber H 1995 *Acta Metall. Mater* **43** 329-337
[5] Ali B, Mohd Sabri M F, Jauhari I and Sukiman N L 2016 *Microelectron. Reliab.* **63** 224-230
[6] Fallahi H, Nurulakmal M S, Fallahi Arezodar A and Abdullah 2012 *J Mater. Sci. Eng.* **A 533** 22-31
[7] Kotadia H R, Mokhtari O, Clode M P, Green M A and Mannan S H 2012 *J Alloys and Compds.* 511 176-188
[8] Kanlayasiri K, Mongkolwongrojn M and Tadashi A 2009 *J Alloys and Compds.* **485** 225-230
[9] El-Daly A A, Fawzy A and Mansour S F 2013 *J Mater Sci: Mater Electron* **24** 2976-2988
[10] El-Daly A A, Desoky W M, Elmosalami T A, El-Sharawy M G and Abdraboh A M 2015 *Materials & Design* **65** 1196-1204
[11] Ma D and Wu P 2016 *Mater. Sci. Eng.* **A 625** 499-506
[12] Liu X D, Han Y D, Jing H Y, Wei J and Xu L Y 2013 *Mater. Sci. Eng.* **A 562** 25-32
[13] Xu L Y, Wang L X, Jing H Y, Liu X D and Wei J 2015 *J Alloys and Compds.* **650** 475-481
[14] Mayappan R, Yahya I, Ghani N A A and Hamid H A 2014 *J Mater Sci: Mater Electron* **25** 2913-2922
[15] Shobby M, El-Refai A M and Fawzy A 2016 *J Mater Sci: Mater Electron* **27** 2349-2359
[16] Mayappan R, Ismail A B, Ahmad Z A, Ariga T and Hussain L B 2007 *J Alloys and Compds* **436** 112-117

Mechanical Behavior of PBO Fiber Used for Lunar Soil Sampler

Xingwen Gao[1], Dewei Tang[1], Honghao Yue[1], Fei Qiao[1] and Yanwei Li[2]

[1]Research Center of Aerospace Mechanism and Control, Harbin Institute of Technology, Harbin, 150080, China;
[2]School of Chemistry and Chemical Engineering, Harbin Institute of Technology, Harbin, 150080, China.
E-mail: gaoxing_hit@163.com

Abstract. The stability of the mechanical properties of the materials used for lunar soil sampler at different temperatures is one of the key factors to ensure the success of the lunar sampling task. In this paper, two kinds of poly(pphenylene-2,6-benzobisoxazole) (PBO) fiber fabric used for lunar soil sampler, flexible tube and wireline, are tested for mechanical properties. The results show that the mechanical properties of the PBO flexible tube and wireline raised 8.3% and 5.7% respectively in -194℃ environment comparing with the room temperature of 25℃. When the temperature rises to 300℃, the deviation is -38.6% and -46.4% respectively.

1. Introduction

PBO fiber perform excellently in specific strength, high specific modulus and temperature tolerance, thus it is considered as a new generation of composite materials in aerospace [1]. The physical and chemical properties of PBO fiber keep stable in the -200℃ ~ +300℃ temperature range, thermal decomposition temperature up to 650℃, with the highest thermal stability in synthetic fiber[2-4]. A diameter of 1mm PBO filament can lift 450kg of weight, that the tensile strength of the fiber is more than 10 times of steel fiber. High modulus PBO fiber have a dimensional stability that the plastic deformation does not exceed 0.03% at 50% fracture load lasting 100h[5-7]. At the same time, the creep resistance of PBO fibers under 5% fracture load is twice that of para-aramid fiber under the same conditions. PBO fiber also has a negative linear expansion coefficient, with a thermal expansion coefficient of -3.0×10^{-6}[8,9]. Its size will not change due to humidity changes, and the impact of heat and moisture is minimal. The friction coefficient of PBO fiber and stainless steel is 0.2, and it is good wearable material. The softness of PBO fiber is excellent, similar to polyester fabric. This characteristic is extremely beneficial for textile weaving process[10]. The flexible tube and wireline used for lunar soil sampler are woven with PBO fibers, as shown in Figure 1. The designed maximum load capacity is 600N.

a)Flexible tube b)Wireline

Figure.1. PBO fiber fabric

2. Experimental design

2.1. Experimental methods

In order to simulate the lunar environment, three sets of comparative tests were carried out on flexible tube and wireline. The sets of flexible tube and wireline are treated for 3 hours at temperature of 300 ℃, -194 ℃ and 25℃ respectively. Then they are tested at room temperature on the tensile machine.

2.2. Experimental apparatus and parameters

As shown in Figure 2a, an electronic universal tensile testing machine, with maximum load of 50KN, beam speed range 0.02-500 mm/min, and load accuracy ± 0.5%, can be used in load, displacement, strain three control methods; Figure 2b shows a liquid nitrogen cryostat using liquid nitrogen cooling in the standard atmospheric pressure, and its minimum temperature can reach -195.8 ℃; Figure 2c shows an electric drying oven, that the control temperature ranges in 10-300 ℃.

a) electronic universal tensile testing machine b) liquid nitrogen cryostat c) electric drying oven

Figure.2. Experimental apparatus

2.3. Experimental process

Using different special fixture, the initial pre-tension of the flexible tubes are set to 100N and 20N, respectively. The quasi-static loading speed is 10mm/min. The test procedure was repeated three times in each group, as shown in Figure 3.

a)Tensile test of flexible tube b)Tensile test of wireline

Figure.3. Tensile Test

3. Experimental design

The failure status of the flexible tube and wireline treated at different temperatures is shown in Fig.4. The woven process of the flexible tube is plain weave, warp and weft in crossing form. This kind of weaving is also the most qualitative, that the organizational structure is not easy to change. Therefore, during the process of tensile failure, the development of the warp fracture is broken one yarn by the other until it is completely destroyed. Yet, the wireline is oblique weave, and the warp or weft yarns will overlap each other with two or more yarns. Therefore, the destruction is instantaneous.

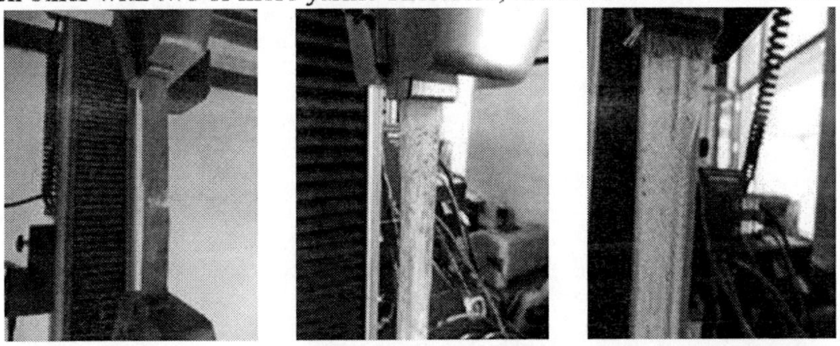

a) b) c)

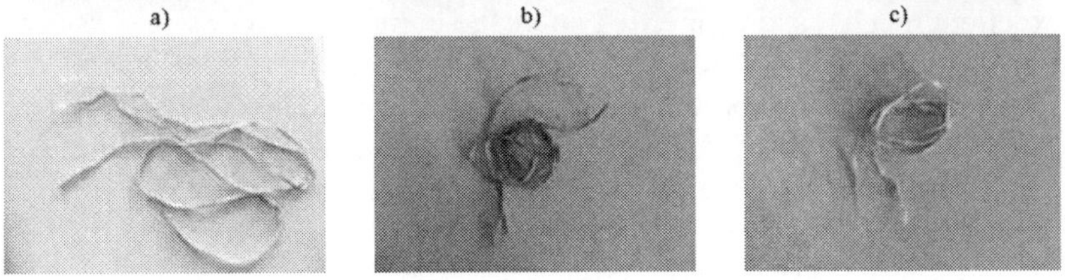

d) e) f)

Figure.4. State of failure

a) Flexible tube at room temperature b) Flexible tube at high temperature c) Flexible tube at low temperature d) wireline at room temperature e) wireline at high temperature f) wireline at low temperature

The width of the flexible tube is 23.6 mm and the thickness is 0.6 mm. The stretching curve is shown in Fig.5. It can be seen that the mechanical properties of flexible tube after cryopreservation have increased by 8.3% compared with that under normal temperature. The mechanical properties of flexible tube after high temperature treatment are lower than that of normal temperature, which is -38.6% lower, but still over the design target of 600N.

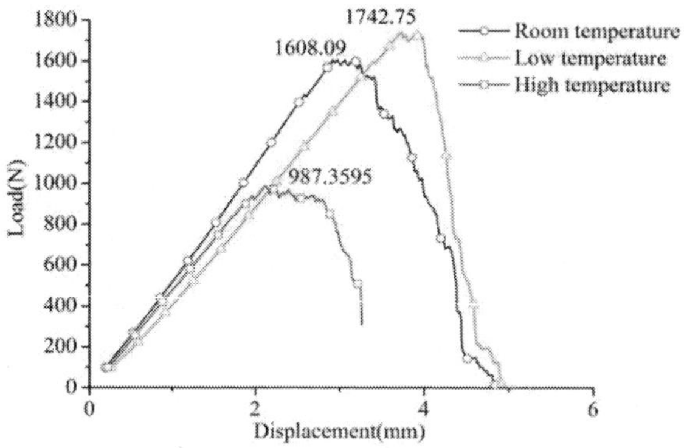

Figure.5. Tensile curve of flexible tube at different temperature

The width of the wireline is 3 mm and the thickness is 0.5 mm. The stretching curve is shown in Fig.6. It can be seen that, similar to the flexible tube, the mechanical properties of wireline after cryogenic treatment have risen 5.7% compared to that under normal temperature. The wireline mechanical properties decreased significantly after high temperature treatment. The amount of change is -46.4% compared to room temperature conditions, still over the design target of 600N.

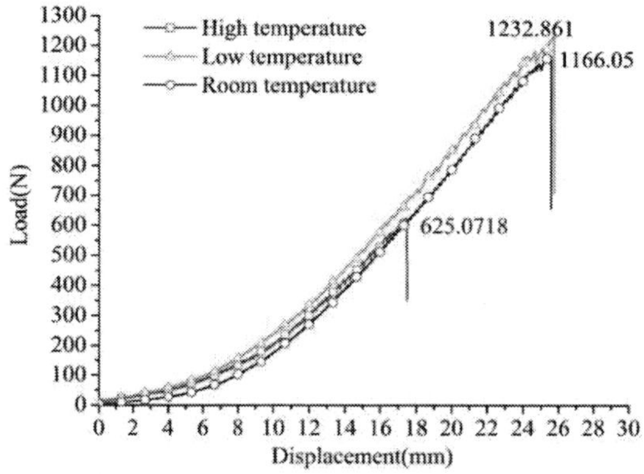

Figure.6. Tensile curve of wireline at different temperature

4. Conclusion

PBO fiber has excellent mechanical properties, after the complex weaving process and surface coating treatment process, the PBO finished products can meet the lunar surface conditions under the load requirements. The mechanical properties of PBO products after the high temperature treatment decreased slightly, the decline rate of flexible tube is -38.6% and that of wireline is -46.4%, while the low temperature treatment will enhance its mechanical properties, the rise rate of flexible tube is 8.3% and that of wireline is 5.7%.

References

[1] Huang Y d 2001 PBO super fiber and its surface treatment *Hi-tech Fiber&Application* **28(1)**:11-16
[2] Hirahata H and Yabuki K 1996 Properties and applications of PBO fibers *Functional Materials* **16(6)**:5
[3] Kitagawa T and Ishitobi M 2000 Analysis of deformation process on poly-p-phenylenebenz-obisoxazole fiber and a structural study of the new high-modulus type PBO HM$^+$ fiber *Journal of Polymer Science Part B: Polymer Physics* **38(12)**: 1 605
[4] Kitagawa T, Yabuki K and Robert J 2002 An investigation into the relationship between processing, structure, and properties for high-modulus PBO fibers. II. Hysteresis of stress-induced Raman band shifts and peak broadening, and skin-core structure *Journal of*

Macromolecular Science Part B **41(1)**:61-76.

[5] Liu D, Zhou X, Zhao Y, et al 2005 Properties and application of PBO fiber *China Synthetic Fiber Industry* **28(5)**:43-46

[6] Wang B Y and Ren P G 2001 The exploriatory study on the advanced PBO fiber composite *Journal of Solid Rocket Technology* **24(4)**:66-71

[7] Joung-man P, Dae-sik K, Sungryong K 2003 Improvement of inter facial adhesion and nondestructive damage evaluation for plasmatreated PBO and Kevlar fibers/epoxy composites using micromechanical techniques and surface wettability *Journal of Colloid and Interface Science* **264(2)**: 431－445.

[8] Larsson F., Svensson L 2001 Carbon, polyethylene and PBO hybrid fibre composites for structural lightweight armour *Composites-Part A: Applied Science and Manufacturing* **33(2)**: 221－231.

[9] Chun-Jie M A, Ning R C 2004 Development and prospect of PBO fiber. *Hi-tech Fiber & Application* **29(3)**:46-51

[10] Li Y 2002 The high performance fiber of 21th century-PBO fiber *Synthetic Fiber in China* **31(1)**:18-22

IOP Conf. Series: Materials Science and Engineering **205** (2017) 012019 doi:10.1088/1757-899X/205/1/012019

Static Strength of Adhesively-bonded Woven Fabric Kenaf Composite Plates

Hilton Ahmad, Lee Sim Yee and Khairi Supar

Faculty of Civil and Environmental Engineering, Universiti Tun Hussein Onn Malaysia, 86400 Parit Raja, Batu Pahat, Johor, MALAYSIA.

hilton@uthm.edu.my

Abstract. Natural fibers are potentially used as reinforcing materials and combined with epoxy resin as matrix system to form a superior specific strength (or stiffness) materials known as composite materials. The advantages of implementing natural fibers such as kenaf fibers are renewable, less hazardous during fabrication and handling process; and relatively cheap compared to synthetic fibers. The aim of current work is to conduct a parametric study on static strength of adhesively bonded woven fabric kenaf composite plates. Fabrication of composite panels were conducted using hand lay-up techniques, with variation of stacking sequence, over-lap length, joint types and lay-up types as identified in testing series. Quasi-static testing was carried out using mechanical testing following code of practice. Load-displacement profiles were analyzed to study its structural response prior to ultimate failures. It was found that cross-ply lay-up demonstrates better static strength compared to quasi-isotropic lay-up counterparts due to larger volume of 0° plies exhibited in cross-ply lay-up. Consequently, larger overlap length gives better joining strength, as expected, however this promotes to weight penalty in the joining structure. Most samples showed failures within adhesive region known as cohesive failure modes, however, few sample demonstrated interface failure. Good correlations of parametric study were found and discussed in the respective section.

1. Introduction

Over the past few decades, there has been a growing interest among material engineers and scientists to develop high-performance engineering materials made from natural fibres to produce composite materials. Natural fibres are potentially used in manufacturing lightweight composites and coupled with lower costs (as compared to synthetics fibres) leading to excellent constituents in engineering materials production. Natural fibers composites also offers advantages such as low tool wear, low density, less hazardous, renewability and biodegradability. Kenaf fibres contain numerous elongated individual fibers which ranging within 6 - 30 μm in diameter and was found that the diameters of the natural fibres has irregular shape and unicircular, demonstrates inconsistency of fibre cross-section along the width. Most researchers assumes natural fibres cross-section as circular shape and average diameter across the fiber width was taken (Roa and Roa, 2007). Munawar *et al.* (2007) determine the cross sectional area of several natural fibres using SEM and obtained an average tensile strength and modulus of 357 MPa and 9.1 GPa, respectively.

CFRP and GFRP are widely used in various sectors as these materials have excellent specific strength to density compared to other engineering materials. Malaysia had limited applications on these materials in engineering sectors as these materials are regarded as imported goods. In civil engineering sector, these materials (particularly CFRP) are commonly used in repair and strengthening works but have limited applications due to high cost. Kenaf fiber is not a new material, it is used in furniture manufacturing industry and paper pulp productions. In Ninth Malaysian Plan (RMK9), government concentrates on biotechnology research to develop aggressive agricultural growth in Malaysia including massive kenaf plantations. The most challenging part in producing natural fiber composites is to combine and provides good bonding between hydrophilic natural fiber and hydrophobic polymer, but with proper treatment, a transparent and good bonding composite material can be obtained (Lee and Ahmad, 2016).

Composite structures components require reliable joining method to assemble composite parts to form a stable structure. Mechanically fastened joint is the most conventional method in structures assembly, but introduction of hole lead to stress concentration, weight penalty, etc. Supar and Ahmad (2016) has reported that complex fracture path were exhibited in different hole configurations of multi-holes GFRP plates. His work were then expended to determine bearing strength at failure of hybrid bolted-bonded joints which were dependent upon various parameters such as applied bolt load, lay-up types, overlap length, and plate thickness (Lee and Ahmad, 2017). Bolted joints and hybrid joints leading to weight penalty and highly stress concentration at the notch tip that may reduce loading capacity of composite plates.

Alternatively, adhesive joint method can be opted but joining strength is largely dependent upon variables such as adhesives and adherends strength, adhesive thickness, overlap length etc. Among the drawback in implementing later method are it requires surface preparations, large peel stresses at far edge, etc. Joint type chosen (i.e., single-lap or double-lap joints) also attributes to different phenomenon in structures response. There is lacking of literatures reported to study the behaviour of any natural fiber composite material subjected to tensile loading. Experimental works were carried out to study static strength of adhesive joint in woven fabric kenaf plates of as a function of identified variables.

2. Experimental Set-Up

2.1 Fabrication of Composite Panels

Fabrication is conducted in the textile laboratory, Universiti Tun Hussein Onn Malaysia (UTHM). A mixture SP84 epoxy and SP76 hardener were used as matrix binder with a ratio of 2:1 respectively. Thin silicon was applied on the formwork to provide smooth surface of finishing plate. A layer of woven fabric kenaf fibre was laid on fabrication aluminium mould and epoxy matrix were poured and smeared carefully to remove the void and air from laid woven fabric layers. The process is then carefully repeated to overlaid woven fabric layers according to stacking sequence desired. Most thermoset systems have one fixed mixing ratio that must not be varied if the optimum cured properties need to be obtained. Then, the fabrication panels were allowed to compress under pressure in order to provide the uniform surface using hydraulic compression machine in 24 hours. The fabricated panels were sectioned with a slow cutting speed water-cooled diamond saw to produce required number of testing coupons. The coupons are numbered and labelled accordingly to the respective series using a designated code.

2.2 Testing Series and Mechanical Testing

There are two different woven fabric architecture types investigated in current study, two sets of plain weave cross-ply and quasi-isotropic KFRP respectively as given in Table 1. The nominal thickness of laminated one layer and two layer kenaf fabric is approximately 2.4 mm and 4.8 mm respectively.

Table 1. The requirement of the sample for the testing

Lay-up	Joint type	Overlap length	Adhesive thickness (mm)
$(0/90)_s$ $(0/90)_{2s}$	Single & Double Lap	10,20	0.2
(0/90/+45/-45) (0/90/+60/-60)	Single & Double Lap	10,20	0.2

Two types of joint types were investigated as shown in Figure 1, i.e., single-lap bonded joint and double-lap bonded joint. Testing coupon with overlap length of 10 mm and 20 mm has nominal gauge length of 145 mm and 150 mm respectively. The end-tab is provided to give better gripping with the testing machine to avoid any slipping during testing. Surface preparations at the end-tab are required by polishing the edge laminate with sandpaper to give good bonding with attached spacer.

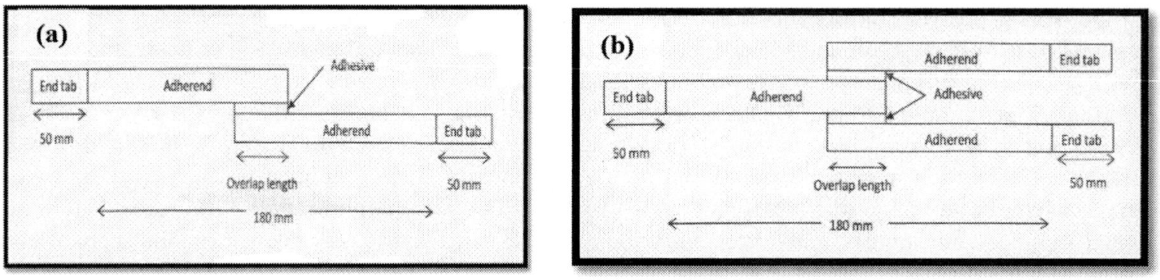

Figure 1. The geometry and dimension of (a) adhesively single-lap bonded joint (b) adhesively double-lap bonded joint.

Two-component of epoxy adhesive Araldite was coated over the lapped region with adhesive thickness of approximately 0.2 mm and allowed to cure for 3 hours at room temperature to reach maximum strength as recommended by the manufacturer. The surface area of the joining adherend is treated with sufficient roughness and a fillet is provided in order to reduce stress concentration due to exhibited peel stress at the plate edge as tensile load applied. The mechanical tests were carried out using Instron 1175 Universal Testing machine (UTM) machine 3369. A crosshead speed of 0.5 mm/min, and a load cell of 100 kN was used. Load and strain data were recorded at one-second intervals using a PC data-logging package from Instron. The ASTM standard D 3039 was followed wherever possible in determining the in-plane tensile properties of the laminates investigated. A minimum of three testing coupons for each joint configuration were tested and demonstrated good reproducibility.

3. Results and Discussions

3.1 Load-displacement Profiles

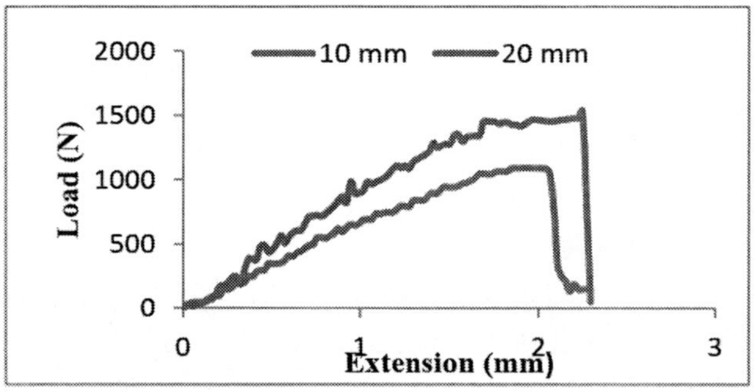

Figure.2. Load-displacement of 10 mm and 20 mm overlap length of single-lap joint of $(0/90)_{2s}$

From Figure 2 above, bonded joint demonstrates fairly linear behaviour from initial phase to ultimate failure and failed immediately as has it reached its maximum load. Woven fabric kenaf composite plate will not experienced any plastic deformation as subjected to tensile loading composite material and considered to fail in brittle manner. Prior to the ultimate failure, it was observed that the coupons were started to crack within the adhesive region. A screeching crack sound of brittle failure was heard prior to joint failure. It is observed that single-lap joint with 20 mm overlap length can withstand a larger peak load compared to 10 mm overlap length counterparts. As a failure are predominantly occurred within the adhesive layer, it is expected that larger overlap joint gives better joint strength. For the single-lap joint of 10 mm and 20 mm overlap, the failure mode occur is regarded as cohesive failure. Cohesion bond failures result in fracture of the adhesive and are characterised by the clear presence of adhesive material on the matching faces of both adherends. Failure is usually by shear, but peel stresses or a combination of shear and peel may also cause a cohesion failure.

3.2 Static strength of adhesive joints of all configurations

Table 2. The maximum load at failure at different variables

Lay-up	Maximum load at failure, P_{max}			
	Single-lap		Double-lap	
	10 mm overlap	20 mm overlap	10 mm overlap	20 mm overlap
$(0/90)_s$	977.48	1213.5	1231.25	1703.1
$(0/90)_{2s}$	1110.85	1607.85	1332.37	2167.8
(0/90/+45/-45)	787.53	956.8	1053.06	1225.25
(0/90/+60/-60)	912.62	1172.3	1186.19	1562.2

Table 2 shows the maximum loads at failure in all testing configurations as investigated in testing series. The largest peak load is shown in double-lap cross-ply $(0/90)_{2s}$ lay-up with 20 mm overlap length while the least value of joint load demonstrated by quasi-isotropic (0/90/+45/-45) lay-up with 10 mm overlap length. Even though double-lap joints leading to weight penalty, but some joint design requires more lapping joint especially in tension joint configurations.

3.1.1 Static strength of adhesive joints as a function of joint type. The single-lap joint is the most common joint used mainly due to its simplicity and efficiency. However, major problem associated to this joint type is the fact that the higher stress concentration (shear and peel) is concentrated at overlap far edge. On the other hand, stress concentration has been reduced in double-lap joints due to the applied load were transferred symmetrically to the loading path. Double-lap joint configuration of $(0/90)_{2s}$ has the largest strength while the least strength is single-lap joint of $(0/90)_s$. This is well understood as single-lap joints exhibits secondary bending that may decreased its carrying capacity compared with double-lap joint counterparts. Similar behaviours are also reported by Ahmad et al. (2014) on bolted joint study. Although double-lap joint exhibited lower peak load, however due to design constraint and client requirement, single-lap joint is unavoidable in structures assembly.

3.1.2 Static strength of adhesive joints as a function of overlap length. As loading were applied, plastic deformations in adherends took place when the strength of bonded joints is larger than proportional limit of adherend. In small overlapping length, bonded area is reduced. Joint is exposed to shear stresses induced as adherend deforms and leading to smaller bending stress. With increasing applied load, adhesive stresses can overcome the elastic limit values in adhesive/adherend interface. Therefore failure initiation and crack propagation occurred within overlapped region. By increasing overlap length, stresses through the overlapping region are decreased. As overlap length increased, joint strength increases due to larger bonding area. The larger overlap length will cause the crack propagation become slower and sustained more load than smaller overlap joints counterparts.

Although larger overlap length contributes to higher joint load, but it may lead to weight penalty and architectural unpleasant. Depending on design requirements, shorter overlap length is more favourable.

3.1.3 Static strength of adhesive joints as a function of lay-up types. It was apparently that adherend (composite) plate with larger volume of 0° plies can bear higher joint load, as expected. This is because the tensile load is applied longitudinally with joining adherend and tensile stress is therefore captured by 0° plies. Bear in mind that woven fabric layer comprised of equivalent two uni-directional layers, each woven fabric layer comprises of two sets or orthogonally unidirectional layer with crimping region. Equal numbers of woven fabric layer in composite plate showed that cross-ply lay-up has larger volume fraction of 0° plies than quasi-isotropic lay-up counterparts. The fiber orientation has a prominent influence on the joint strength to resist applied load where arrangement of fiber direction (0°) parallel to applied loading direction give optimum strength. When the surface fiber direction coincides with the loading direction, the strain on the surface of the laminate is limited due to the presence of the reinforcing fibers. From the experimental findings, it shows that the cross-ply lay-up can sustain higher load than equivalent quasi-isotropic lay-up. Since the quasi-isotropic lay-up has low volume fraction of 0° ply in loading direction, it failed at relatively lower load.

Conclusions

Testing series and associated effects were investigated and careful measurements have been made. It was found that by increasing overlap length promotes to better joint strength. However, from material design perspective, large overlap length contributes to weight penalty and this reduce the advantages of applying composite material that offers excellent strength to density ratio. Adhesive joints configuration and lay-up types also gives significant contribution to the tensile load applied during mechanical testing. Secondary bending occurred in single-lap joint that alter the stress distribution profile and significantly reduce the single-lap joint strength compared to double-lap joint counterparts. This explains less good joint strength in single-lap joint configurations. Cross-ply lay-up demonstrated better joint strength than equivalent quasi-isotropic lay-up. It is suggested that due to higher volume fraction of 0° plies in cross-ply lay-up offer better resistance to applied tensile load.

Acknowledgements

The author would like to thank Universiti Tun Hussein Onn Malaysia (Vot. U245) in providing financial support for this project work.

References

[1] Roa K.M.M and Roa K.M (2007). "Extraction and tensile properties of natural fibres:Vakka, Date and Bamboo", Composite Structures, Vol. 77 , pp. 288 – 95
[2] Munawar S.S., Umemura K, and Kawai S. (2007). Characterization of the Morphological, Physical and Mechanical Properties of Seven Nonwood Plant Fiber Bundels. Journal of wood Science, Vol. 53, pp. 108-113.
[3] Ahmad H, Crocombe A.D., and Smith P.A. (2014) "Strength prediction in CFRP woven laminate bolted single-lap joints under quasi-static loading using XFEM." Composites Part A, Vol. 66, pp.82–93.
[4] Lee S.Y.and Ahmad H. (2016). "Experimental Strength of Single-Lap Hybrid Joints on Woven Fabric Kenaf Fiber Composites Under Quasi Static Condition", MATEC Web of Conferences, Vol. 47, pp. 1-6.
[5] Supar K. and Ahmad H. (2017). "XFEM Modelling of Multi-holes Plate with Single-row and Staggered Holes Configurations", MATEC Web of Conferences, Vol. 103, pp. 1-8.
[6] Lee S.Y.and Ahmad H. (2017). "Bearing Stress at Failure of Double-Lap Hybrid Joints in Woven Fabric Kenaf Fiber Composite Plates under Quasi-static Loading", MATEC Web of Conferences, Vol. 103, pp. 1-7.

Comparison of Energy Absorption of Aluminium-composite Tubes Subjected to Axial Loading

Tahir Abbas[1], H.H.Ya[2], Mohamad Zaki Abdullah[3]

[1,2,3]Mechanical Engineering Department Universiti Teknologi PETRONAS 32610 Seri Iskandar, Perak Darul Ridzuan, Malaysia
alitahir739@gmail.com[1]
hamdan.ya@petronas.com.my[2]
abdullah@petronas.com.my[3]

Abstract. In this paper, energy absorption capability and failure modes of partially wrapped aluminium tubes have been studied. These tubes are used in the front side of automobiles and aircraft applications. Filament winding technique was used for partial wrapping of these tubes. Partially wrapping on the external surface of aluminium tubes was done with glass fibers, and epoxy resin, which is a composite material. Various composite layers and fiber angles were used in partial wrapping, which includes 4, 6 and 8 composite layers of $\pm$ 55° fiber angle. These tubes were subjected to axial crushing using the universal testing machine, and testing speed was 5mm/min. Failure modes and energy absorption analysis were carried out after testing. The experimental results revealed that partially wrapped aluminium tubes are 3.81%, 8.13% and 17.06% more efficient in energy absorption as compared to the tubes without wrapping. Furthermore, the effect of composite layers and failure modes has also been described.

1. Introduction

Nowadays, light-weight engineering materials are the best choice for reduction of vehicles weight as well as to deal with challenges of everyday higher prices of fuel [1-3]. Aluminium is one of light-weight engineering material that has plenty of applications in automobiles. One of the important applications of aluminium tubes is on the front side of vehicle structure to absorb impact. Moreover, their ease of fabrication, low cost, and low density make them more attractive for such applications [4]. Moreover, a combination of metal and composite material has been used to improve crashworthiness properties. This hybrid combination of material takes part to enhance impact properties of tubes during crash. Because high strength to weight ratio of composite material contributes to absorbing maximum energy absorption and it also takes part to change failure mode of these tubes [5]. During axial crushing, plastic deformation of metal occurs while composite material fails due to multiple fracturing of fibers, these both phenomena contribute energy absorption in these tubes [6]. Furthermore, axial and hoop strength of composite would also effect the collapse modes and energy absorption [7]. In this paper, an effort has been made to improve energy absorption capability of aluminium tube after partial wrapping of it with glass/epoxy. The basic objective of partial wrapping is to enhance its impact properties as well as to achieve weight reduction. These tubes are tested using the universal testing machine and impact properties have been analyzed in detail.

2. Literature Review

IOP Conf. Series: Materials Science and Engineering **205** (2017) 012020 doi:10.1088/1757-899X/205/1/012020

Bi-material tubes were first, studied by Mamalis [8]. Various factors like, composite and metal thickness, material properties, and geometrical construction were identified that contribute to energy absorption. These factors were documented as the key parameters for energy absorption.
Energy absorption of square cross-section aluminium tubes, externally wrapped with glass fiber reinforced polymer, were investigated by Shin [9]. These tubes were tested under axial compression. To analyze the effect of fiber orientation on energy absorption aluminium tubes, they were wrapped with three different fiber orientations, 0°, 90° and ± 45°. The process selected to develop test samples was filament winding, and prepreg sheets of glass/epoxy were used to fully wrap aluminium tubes. The most important relevant finding of this study was, aluminium tube wrapped with 90° fiber orientation indicated more energy absorption as compared to other tubes. Research on energy absorption capability of steel tubes of the hollow cross-section was published by Bambach [5, 10]. Samples were prepared according to the procedure adopted by Shin [6]. But, instead of glass/epoxy, carbon fiber reinforced polymer prepreg sheets were used for external wrapping of steel tubes. Samples were tested under static and dynamic loading. The results of this study indicated that tubes of 90° wrapping were good to hold the maximum load. It was further noted that tubes with 90° wrapping absorbed more energy. The study of the effect of composite layers on steel tubes was carried out by Kalhor [6]. The process used to prepare the sample was filament winding, and wrapping was done using fiber-reinforced plastics. In this study, samples were fully wrapped as previously done by Bambach. This study set out with the aim of assessing the effect of composite layers on energy absorption. The most obvious finding of this study was, as composite layers increase energy absorption increases. Wet filament winding process was used by Song [4], for external wrapping of steel tube samples. Three different fiber angles of glass/epoxy, for wrapping of tubes, were selected to see their effect on specific energy absorption (SEA) as well as collapse modes of tubes. The results of this research revealed that 90° fiber orientation wrapped tubes absorbed more energy. This experimental study also detected that with increasing number of layers of composite material, and fiber orientation specific energy absorption also increases. Energy absorption capability of both circular and square cross-section aluminium tubes was examined by Babbage [11]. Samples were prepared by the same process as adopted by Song [4]. The material used was glass/epoxy, and two fiber orientations, ± 45°and ± 75° were selected. The results revealed that circular cross-section tubes absorb more energy as compared to square tube. Additionally, ± 75° wrapped tube absorb more specific energy for both cross-sections. More recently, a study has been conducted by Lima [12], on partially wrapped steel tubes, to examine their energy absorption. Wet filament winding process was used to develop samples. Three fiber orientations ± 45°, ± 75° and 90° were used, for three different composite layers 4, 6, and 8. Testing was carried out under quasi-static loading. It was reported that 90° wrapped tube with 6 number of composite layers absorbed more energy. In view of all that has been mentioned so far, it is concluded that there is need to study energy absorption capability of partially wrapped aluminium tubes as well. In this paper an attempt has been made, to study crashworthiness properties of partially wrapped aluminium tubes. Glass/epoxy composite material has been selected due to its lower price as well as good energy absorption ability as shown in Fig 1[13]. Moreover, wet filament winding process has been used to develop samples. Furthermore, the sample is tested using the universal testing machine.

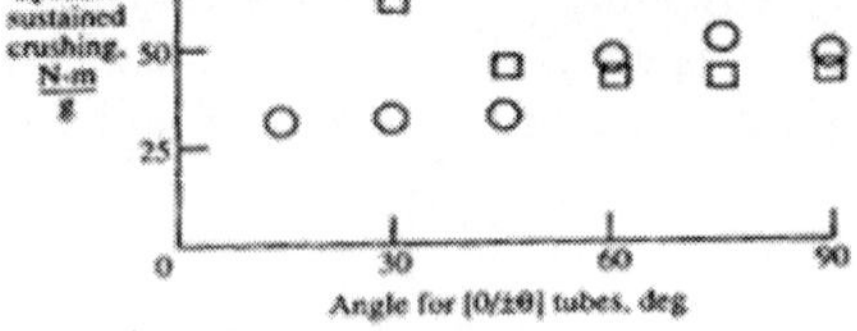

Figure.1. Energy absorption ability of composite tubes containing glass- carbon-epoxy combination [13]

3. Material and Methods

The 'AA6063' grade of aluminium tube was used, which is commercially available. Glass fiber and epoxy resin grades that were used in this research work are E-glass 1200tex, and D.E.R. ™ 331™ respectively. These both grades are also commercially available. To avoid Euler buckling, the length of tubes was used 200mm; that was also suitable for proper collapse mod. Dimensions of selected tubes are given in Figure 2 and Figure 3. Additionally, Partial wrapping of aluminium tube samples, using 4, 6 and 8 composite layers of ±55° fiber angles, was carried out. The process used for partial wrapping was wet filament. The universal testing machine was used to carry out axial compression, and speed was maintained 0.08 mm/sec (5mm/min), to avoid strain rate effect. Crashworthiness parameters were obtained from Load-displacement (L-D) curves and were analyzed to calculate specific energy absorption and crush force efficiency.

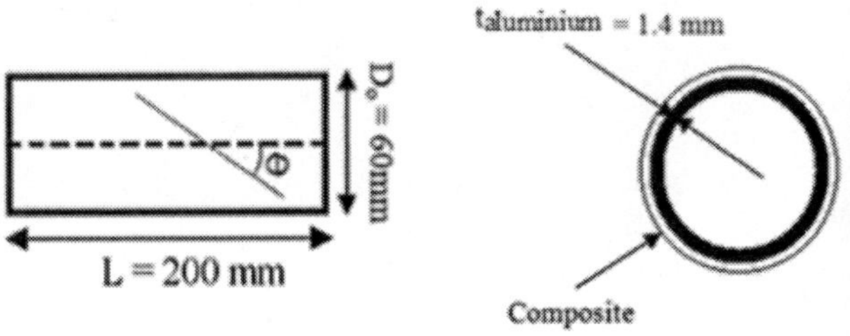

Figure 2. Dimensions of selected aluminium tube

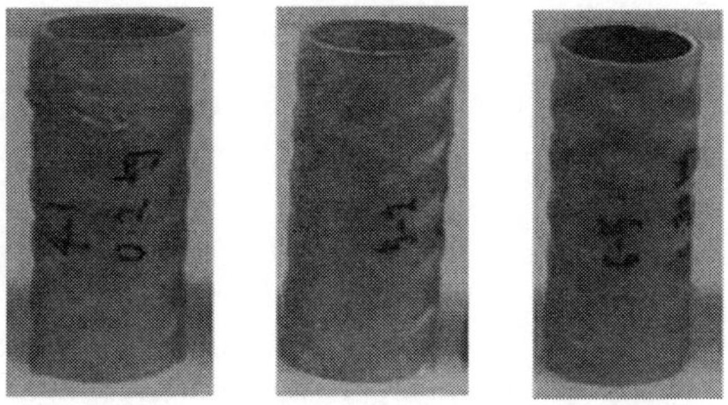

Figure 3. Partially wrapped aluminium tube samples (a) 4 layers (b) 6 layers (c) 8 layers

3.1 Energy absorption parameters

Energy absorption parameters were calculated by using (1) to (4) [14].

$$TEA= \int P_{mean}\, Ds \equiv P_{mean}\,(Df-Di) \quad (1)$$

$$P_{mean}=TEA/\,(Df-Di) \quad (2)$$

$$SEA=TEA/m \quad (3)$$

$$CFE=P_{mean}\,/\,P_{max} \quad (4)$$

Where Df and Di are final and initial crushing displacement and m represent the mass of uniform sample.

4. Results and Discussions

Equations that are presented in section-III used to calculate parameters, that are total energy absorption (TEA), Specific energy absorption (SEA), crush force efficiency (CFE). Maximum load (P_{max}) and average load (P_{mean}) are obtained from load displacement graph.

4.1 Failure modes of partially wrapped and without wrapped aluminium tubes

In Fig 4 there are clear increasing and decreasing trend oscillations of aluminium tubes deformation, subjected to axial compression. Basically, these increasing and decreasing trend lines are depicting rise and fall of the load. At the initial point, load rises which are indication of fold formation this continues until that point after which tube wall started to fold outside. At this stage, load fall rapidly, after a sharp decrease of load there is also an increasing trend which shows internal buckling of the tube. This pattern of maximum and minimum load values continue to repeat till the desired length. Furthermore, Asymmetric concertina failure modes are exhibited by all these tubes as shown in Fig 5 and Fig 6.

Moreover, it is apparent from Table I that specific energy absorption and crush force efficiencies of partially wrapped aluminium tubes have been improved, as compare to aluminium tubes as received. It is further quite clear, that SEA of a partially wrapped tube having 8 composite layers is higher than all other tubes. It is attributed to more number of composite layers. Fiber kinking and delamination are also perceived in tubes with 4 and 6 composite layers. Additionally, crush force efficiency of the partially wrapped tube with 4 composite layers found to be more as compared to other. This is due to the ratio of maximum and minimum load value. This is shown in Fig 7.

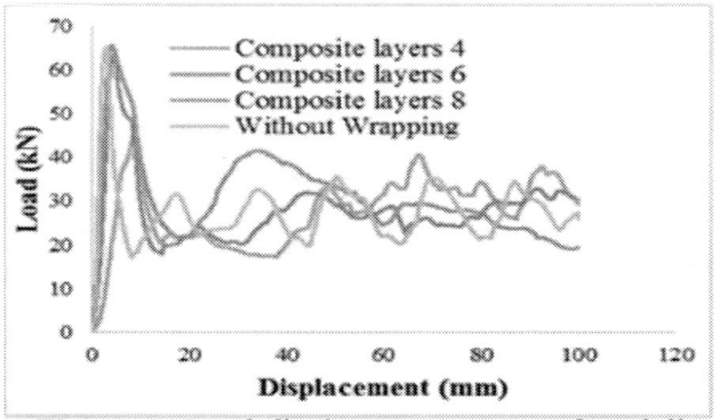

Figure 4. Load-displacement curves of partially wrapped and unwrapped aluminium tubes under axial load

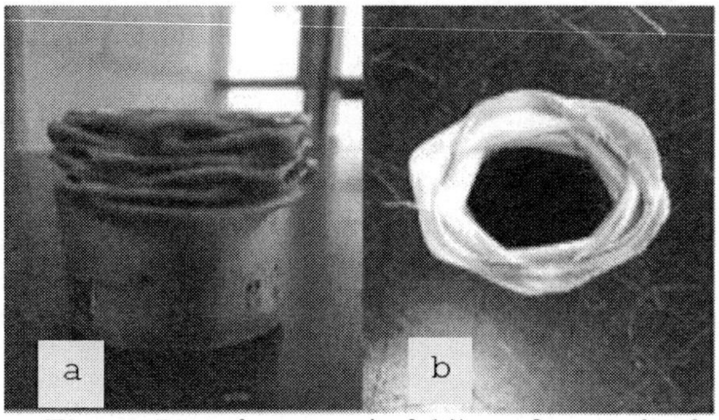

Figure 5. Axisymmetric folding of as received aluminium tube (a) front view (b) top view

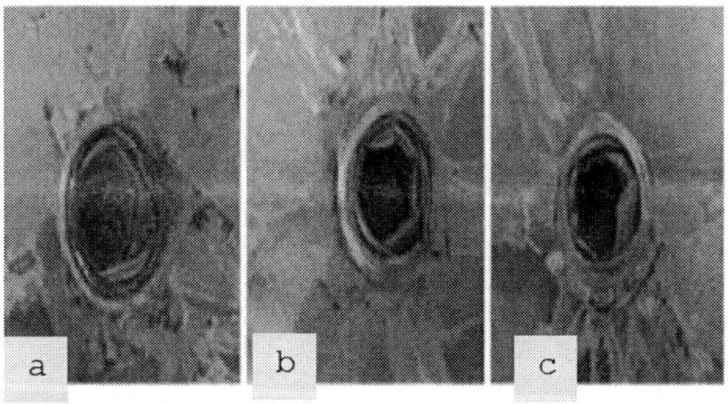

Figure 6. Failure modes of partially wrapped aluminium tubes (a) Failure mode of 4 layers wrapped aluminium tube (b) Failure mode of 6 layers wrapped aluminium tube (c) Failure mode of 8 layers wrapped aluminium tube

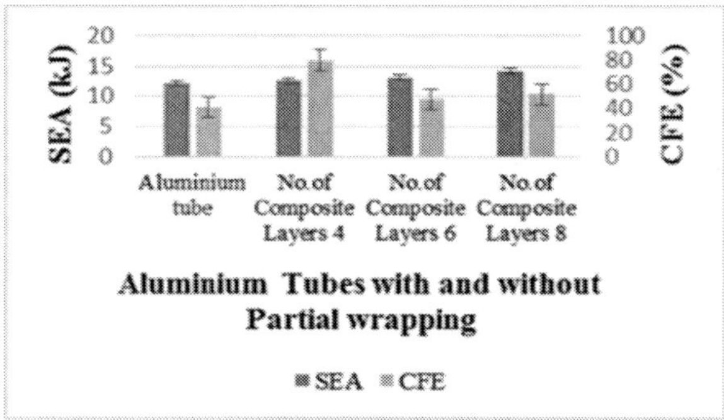

Figure 7. Specific energy absorption and crush force efficiency comparison of partially wrapped aluminium tubes subjected to axial compression

Table 1. . Parameters of energy absorption

Parameters	unwrapped aluminium tubes	Partially wrapped tube with 4 composite layers	Partially wrapped tube with 6 composite layers	Partially wrapped tube with 8 composite layers
TEA (kJ)	2.67	2.78	3.02	3.41
SEA (kJ/kg)	12.13	12.6	13.13	14.20
P_{max} (kN)	65	45.45	63.45	65.80
P_{mean} (kN)	26.7	27.8	30.21	34.03
CFE (%)	41.07	80.16	47.61	51.71

5. Conclusions

This study mainly described the effect of composite layers on energy absorption capability of aluminium tubes. To improve crashworthiness parameters, an effort has been made by the partial wrapping of aluminium tubes with glass/epoxy composite. A comparison of crashworthiness parameters has been listed between as received and partially wrapped aluminium tubes. The results of this study clearly described that wrapped tubes are good in energy absorption. Moreover, specific energy absorption (SEA) of as received aluminium tubes has been improved to 3.81%, 8.13% and 17.06% by the external partial wrapping of them, using 4, 6 and 8 composite layers respectively. In addition to this, to see the effect of composite layers, crashworthiness parameters of partially wrapped tubes are also documented. It is examined, after axial compression, all these tubes exhibited an axisymmetric mode of failures. Additionally, it has been found that aluminium tube wrapped with 8 composite layers absorb more specific energy as compared to other tubes, this is due to more fibers participation during the crushing process to hold the applied load. Despite these promising results, further research should be undertaken to investigate the effect of gaps between composite layers and to find out a best possible combination of fiber angle, composite layer and gap between composite layers.

6. References

[1] R. Kalhor, H. Akbarshahi, and S. W. Case, "Numerical modeling of the effects of FRP thickness and stacking sequence on energy absorption of metal–FRP square tubes," *Composite Structures,* vol. **147**, pp. 231-246, 2016.

[2] Abbas, T., H. Ya, and M.Z. Abdullah, Energy absorption capability of aluminium tube partially wrapped with glass/epoxy subjected to quasi-static loading. ARPN Journal of Engineering and Applied Sciences, vol.**11**, NO. 24, DECEMBER 2016.

[3] Abbas, T., H.H. Ya, and M. Zaki, *The effect of fiber angles on energy absorption of 50% partially wrapped aluminium - composite hybrid tube under axial crushing.* Materialwissenschaft und Werkstofftechnik, 2017.

[4] H.W. Song, Z.M. Wan, Z.M. Xie, and X.W. Du, "Axial impact behavior and energy absorption efficiency of composite wrapped metal tubes," *International Journal of Impact Engineering,* vol. **24**, pp. 385-401, 2000.

[5] M. Bambach and M. Elchalakani, "Plastic mechanism analysis of steel SHS strengthened with CFRP under large axial deformation," *Thin-walled structures,* vol. **45**, pp. 159-170, 2007.

[6] R. Kalhor and S. W. Case, "The effect of FRP thickness on energy absorption of metal-FRP square tubes subjected to axial compressive loading," *Composite Structures,* vol. **130**, pp. 44-50, 2015.

[7] H.H. Ya, H. El-Sobky, "Collapsed Mode and Specific Energy Absorption of Glass Filled Polypropylene (GPE) and Glass Filled Polyethylene (GPP) Composite Pipes under Quasi-Static Axial Loading", Key Engineering Materials, Vol. **673**, pp. 141-149, 2016

[8] A. Mamalis, Y. Yuan, and G. Viegelahn, "Collapse of thin–wall composite sections subjected to high-speed axial loading," *International journal of vehicle design,* vol. **13**, pp. 564-579, 1992.

[9] K. C. Shin, J. J. Lee, K. H. Kim, M. C. Song, and J. S. Huh, "Axial crush and bending collapse of an aluminum/GFRP hybrid square tube and its energy absorption capability," *Composite structures,* vol. **57**, pp. 279-287, 2002.

[10] M. Bambach, M. Elchalakani, and X.-L. Zhao, "Composite steel–CFRP SHS tubes under axial impact," *Composite Structures,* vol. **87**, pp. 282-292, 2009.

[11] J. Babbage and P. Mallick, "Static axial crush performance of unfilled and foam-filled aluminum–composite hybrid tubes," *Composite structures,* vol. **70**, pp. 177-184, 2005.

[12] R. M. Lima, Z. Ismarrubie, E. Zainudin, and S. Tang, "Energy Absorption Capability of Hybrid Tube Made by Mild Steel and GFRP under Quasi-Static Loading," in *Advanced Materials Research*, 2012, pp. 2741-2746.

[13] G. L. Farley, Relationship *between mechanical-property and energy-absorption trends for composite tubes* vol. **3284**: National Aeronautics and Space Administration, Office of Management, Scientific and Technical Information Program, 1992.

[14] Z. Ahmad, "Impact and energy absorption of empty and foam-filled conical tubes," Ph.D. Thesis, Queensland University of Technology, 2009.

IOP Conf. Series: Materials Science and Engineering **205** (2017) 012021 doi:10.1088/1757-899X/205/1/012021

Analysis of the Machining Depth over Time in Wet Etching for Fabrication of Invar Thin Film Shadow Mask

G C Moon[1], W K Choi[1], S H Kim[1], S G Choi[1] and E S Lee[2]

[1]School of Mechanical Engineering, Inha University, Incheon 22212, Republic of Korea
[2]Department of Mechanical Engineering, Inha University, Incheon 22212, Republic of Korea
moonk8924@naver.com

Abstract. Invar is made up of 64% iron and 36% nickel. Invar has a characteristic of low thermal expansion coefficient that hardly causes a change of dimension according to temperature change. It is used for shadow mask fabrication by using this feature. Shadow mask is an important element for enhancing the pixels of the display. Therefore, Wet etching is an important machining technique because it is economical and mass-producible to make shadow mask. In this study, the machining depth was compared and analyzed with the machining time by using a 30μm invar thin film. As the machining time increased, the machining depth of the hole increased and a perfect hole array occurred when the machining time was 5 minutes. Through the same processing conditions, a 5.5-inch large-area shadow mask with uniform hole size was fabricated.

1. Introduction

The shadow mask is an important element for enhancing the pixels of the display, the smartphone and the VR, which are regularly drilled with hundreds of small holes of the same size and mounted on the inner surface of the panel. The shadow mask uses the electron beam injected from the electron gun to reach the fluorescent surface of the inner surface of the panel, and the temperature is high. When deformation occurs in a small hole due to high temperature, red, green, and blue fluorescent dots are not accurately deposited. Therefore, an invar consisting of 64% iron and 36% nickel having a low thermal expansion coefficient will be used. Various machining methods are used to fabricate shadow masks. Laser machining has a problem of taper generation and burrs during machining, removal of burrs is in itself a great challenge for producing zero-defect high density pixel counts and Water jet and Electric discharge machining (EDM) clearly them from entering the current state-of-the-art of making mask. [1] Electro chemical machining (ECM) has difficulty in making a rectangular shape and an over cause a dimensional error. [2] which makes it difficult to manufacture accurate shadow masks and takes a long time to process. Recently, many studies have been attempted to improve spray etching technologies for micro-electronics manufacturing. [3-5] There are many advantages in wet etching. There are no burrs on the edges of the parts, while the substrate is not subjected to either heat or physical deformation. In addition, extremely small features can be obtained and the dimensions can be controlled precisely and wet etching is used to fabricate shadow masks, enabling rapid mass production.[6] Wet etching is the most commonly used method for fabricating shadow masks. However, wet etching has a problem that the workability greatly changes according to the processing conditions. In this study, a dry film pattern with a size of 120 × 80μm is prepared through a photomask of a 30μm invar thin film and the time of hole generation is determined by comparing and analyzing the variation of the processing depth with time of the spray type etching. The 5.5-inch large-area shadow mask was fabricated at the determined machining time of 5 minutes by analyzing the change of the machining depth with time.

2. Experimental details

2.1. Photolithography

Figure.1 shows the process of making a pattern on an invar using photoresist. Laminating was carried out using a dry film manufactured by Hitachi Ltd. under the conditions of a laminating temperature of 110℃ and a laminating speed of 1 m/min. After laminating, the holding time was 10 minutes so that the adhesion of the dry film and Invar completely took place. After the holding time, the glass mask was placed on the stepper coated with the dry film and the exposure was performed for a time of 90 seconds.

In the patterned portion of the glass mask, UV light is not received during exposure, and the dry film is not cured and is peeled by reaction with a developing solution of 4 wt%. This yields a dry film pattern with a micropattern of uniform size to produce a shadow mask.

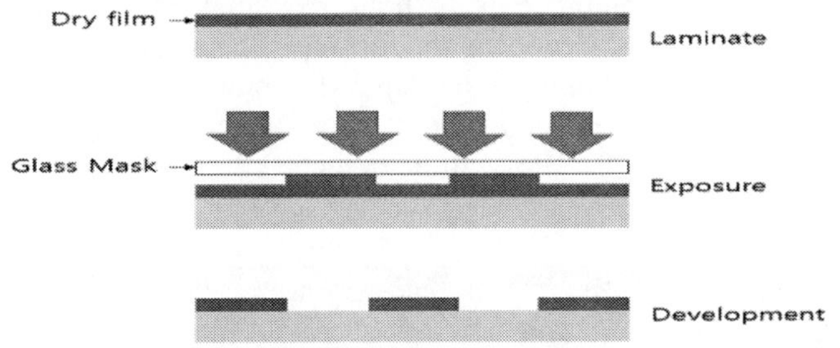

Figure 1. Photolithography process

2.2. Wet etching

Fe and *Ni* constituting the Invar are processed by etching reaction with an aqueous solution of ferric chloride ($FeCl_3$). The Fe^{3+} ions contained in the aqueous solution of nickel chloride are adsorbed while diffusing to the metal surface. The adsorbed Fe^{3+} ion reacts with *Fe* and *Ni* constituting the invar and reacts as shown in equations (1) and (2).

$$2FeCl_3 + Fe \rightarrow 3FeCl_2 \quad (1)$$
$$2FeCl_3 + Ni \rightarrow 2FeCl_2 + \mathrm{NiCl} \quad (2)$$

After the reaction, the ions desorbed from the metal surface are released. Microhole machining is performed as these chemical reactions are repeatedly performed. Figure. 2 is a schematic view of the spray-type wet etching system used in the experiment. Generally, there is a problem that a new $FeCl_3$ aqueous solution must be supplied in order to perform a chemical reaction in the wet etching process. Spray is used to smoothly supply $FeCl_3$ aqueous solution to the invar surface. In addition, by applying pressure, the processing in the depth direction is performed while reducing the isotropic property of the wet etching.

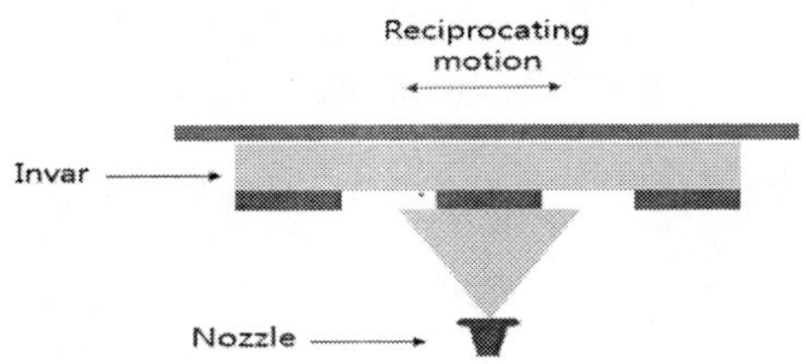

Figure 2. Schematic layout of wet etching system

3. Results and Discussion

When spray etching is performed, the depth of machining is compared with time. By comparing and analyzing the depth of machining, the machining time is determined so that machining is performed by the depth to be machined. Table.1 shows machining conditions for sprat wet etching. We compared the hole size and the depth of hole according to the change of time for the shadow mask production.

Table 1. Experimental machining conditions.

Conditions	Values
Etchant	30wt% $FeCl_3$
Pressure	0.08MPa
Temperature	35℃
Machining Time	1, 2, 3, 4, 5min

Figure. 3 represents the size at the entrance of the hole over time. The size at the entrance of the hole with time increases gradually. This is due to the isotropic characteristics of the wet etching. The isotropic properties are not machined only in the depth direction but in all directions. Because of this isotropic characteristics, when the machining time is too long, it will be affected by the machining between the adjacent holes, and the dry film will be detached and the accuracy of the machined shape will be decreased.

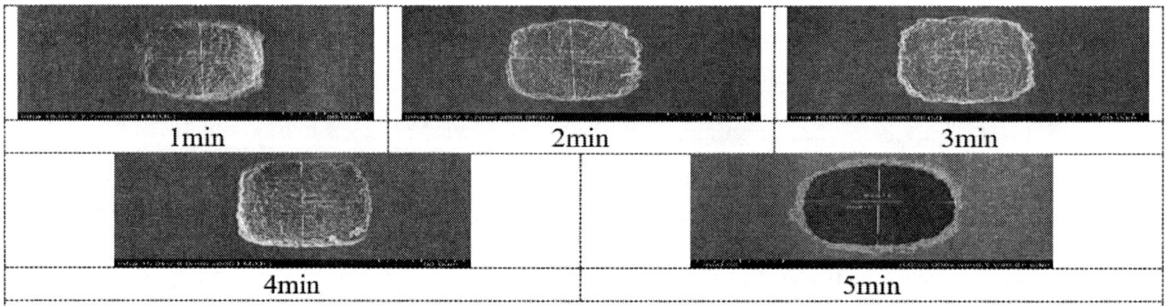

Figure 3. SEM images of machining hole entrance over time

Figure.4 shows the depth of machining over time. The machining depth gradually increases with time. When wet etching is used, the change in machining depth with time is analyzed to determine the time at which holes are generated. The removed invar material in axis direction is the biggest in the first minute of spray wet etching, which is the least contact area of the etchant with the surface of the invar, which is anisotropic rather than isotropic. After 1 min, the depth of machined hole on invar film gradually decreases, because the isotropic characteristic of the wet etch process allows the process to be performed in all directions.

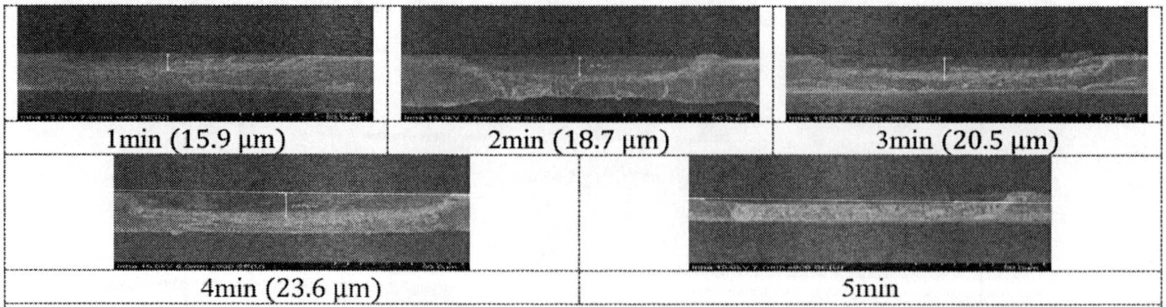

Figure 4. SEM images of machining depth over time

This suggests a high efficiency and good accuracy in creating a micro-hole array by comparing the machining rates of the hole's inlet and processing depth over time. Using a 5-minute machining time to generate a hole array, a 5.5-inch large area shadow mask can be fabricated as shown in Figure 5.

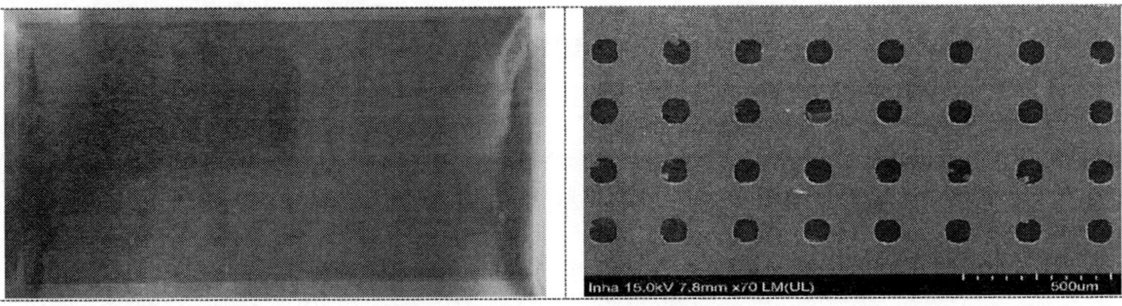

Figure 5. 5.5inch shadow mask

4. Conclusions
1. The inlet of the hole with time becomes larger due to the isotropic nature of the wet etching. The size of the hole gradually increased, but in the 5 minutes that the hole was created, it became 91.6μm in width and 119μm in height.
2. A hole with a thickness of 30μm was produced when the etching time was 5 minutes due to the chemical corrosion reaction of the etchant and the impact force of the spraying pressure.
3. A shadow mask with a uniform size can be generated when the processing time of the processing time 5 minutes in which the hole having the pattern of 120 × 80μm is generated is generated.

5. Acknowledgments
This research was supported by the Strengthening competitiveness business of industrial clusters funded by the Korea Industrial Complex Corporation.

6. References
[1] Sundaram N Kumar, Robin John, Scott Lauer, White Little and Bob Dual 2015 Electroforming Technology for Manufacturing Thin Metal Masks with Very Small Apertures for OLED Display Manufacturing *Symposium Digest of Technical Papers* **46** 211-214

[2] Xiaolei Chen, Ningsong Qu, Hansong Li and Zhengyang Xu 2016 Electrochemical micromachineing of micro-dimple arrays using a polydimethylsiloxane (PDMS) mask *Materials Processing Technology* **229** 102-110.

[3] D S Jin, K H Chun and E S Lee 2016 Analysis of the current density Characteristics in through-mask electrochemical micromachining for fabrication of micro-hole arrays on invar alloy film *Chinese Journal of Aeronautics* (preprint CJA 764)

[4] Geissler M, Schmid H, Michel B and Delamarche E 2003 Selective Wet of Microcontact printed Cu Substrates with Control over the etch Profile *Microelectronic Engineering* **67-68** 326-332

[5] A Visser And M Buhlert 2001 Theoretical and practical aspects of the miniaturization of lead frames by double sided asymmetrical spray etching *Journal of Materials Processing Technology* **115** 108-113

[6] Jeong H C, Kim D J, Choi G M and Kim D J 2009 The Effect of Spray Characteristics on the Etching of Invar Alloy with $FeCl_3$ Solution *International Journal of Precision Engineering and Manufacturing* **10** 107-114,

A Comparative Study of Production of Glass Microspheres by using Thermal Process

May Yan Lee[1,a], **Jully Tan**[1,b], **Jerry YY Heng**[2,c], **Christopher Cheeseman**[3,d]

[1]Faculty of Engineering, Technology and Built Environment UCSI University, No. 1, Jalan Menara Gading, UCSI Height, 56000 Cheras, Kuala Lumpur, Malaysia
[2]Faculty of Engineering, Department of Chemical Engineering, Imperial College London, South Kensington Campus SW7 2AZ, UK
[3]Faculty of Engineering, Department of Civil and Environmental Engineering, Imperial College London, South Kensington Campus SW7 2AZ, UK

[a]my.leeapril94@gmail.com, [b]tanjully@ucsiuniversity.edu.my, [c]jerry.heng@imperial.ac.uk, [d]c.cheeseman@imperial.ac.uk

Abstract Microspheres are spherical particles that can be distinguished into two categories; solid or hollow. Microspheres typical ranges from 1 to 200 µm in diameter. Microsphere are made from glass, ceramic, carbon or plastic depending on applications. Solid glass microsphere is manufactured by direct burning of glass powders while hollow glass microspheres is produced by adding blowing agent to glass powder. This paper presented the production of glass microspheres by using the vertical thermal flame (VTF) process. Pre-treated soda lime glass powder with particle sized range from 90 to 125µm was used in this work. The results showed that glass microspheres produced by two passes through the flame have a more spherical shape as compared with the single pass. Under the Scanning Electron Microscope (SEM), it is observed that there is a morphology changed from uneven surface of glass powders to smooth spherical surface particles. Qualitative analysis for density of the pre-burned and burned particles was performed. Burned particles floats in water while pre-burned particles sank indicated the change of density of the particles. Further improvements of the VTF process in terms of the VTF set-up are required to increase the transformation of glass powders to glass microspheres.

1. Introduction

Microspheres are spherical particles that can be distinguished into two categories; solid or hollow. Microspheres typical ranges from 1 to 200 µm in diameter [1]. Both solid and hollow microspheres can be produced from glass, ceramic, carbon or plastic. Solid glass microspheres are usually made from soda-lime glass due to the low melting point and chemical inertness of soda-lime glass. The conventional method in producing solid glass microspheres is by the In-Flame Spheroidisation Method [2] where a continuous controlled flow of powdered glass is feed to a gas flame. In contrast, hollow glass microspheres are produced by adding a blowing agent to glass powder [3]. Blowing agent such as sodium silicate decomposes to multiple gases when burned, causing the microsphere to form with a hollow structure. Glass microspheres applications as fillers in syntactic foams resulting in reduction of material density, compaction and heat conductivity [4]. Developed Vertical Thermal Flame (VTF) process has potential to produce cenosphere from fly ash with high yields. In the VTF process, the raw materials are fed into a vertical tube via a funnel and the raw material will come in contact with the flame located at the bottom of the vertical tube. The burned particles are collected and cooled in a beaker via a collector plate before proceeding to the particle characterization study [5]. This paper aims to study the potential production of microspheres from glass by using the VTF process.

2. Methodology

2.1. Preparation of feed sample

Recycled glass was used as the primary raw material (Ecosand, supplied by Day Group Ltd, London). This was commercially ball milled to form a powder with a mean particle size of 45μm. Polyvinyl alcohol (PVA) was dissolved in water form a solution. The milled glass was then mixed with a small volume of the binder solution to form spherical glass agglomerate particles with size between 90 to 125μm. PVA is an organic compound that acts as a binder and blowing agent to aid in formation of hollow glass microsphere. When the agglomerated powdered soda-lime glass is burned, the PVA decomposes and resulted in the gaseous evolution of acetic acid, water and hydrogen bromide. The gases evolved within the feed material structure develops the hollow cavity in the glass microspheres.

2.2. Burning Process

The VTF process was done by setting up a vertical blue flame by adjusting the propane gas-to-air ratio. Pre-treated soda-lime glass powders were fed into the quartz tube with a different dimension - length 35cm and width 20mm via a stainless steel filter funnel. The burned particles were collected at the bottom using a collector plate [5]. The particles undergoes second pass for comparison study.

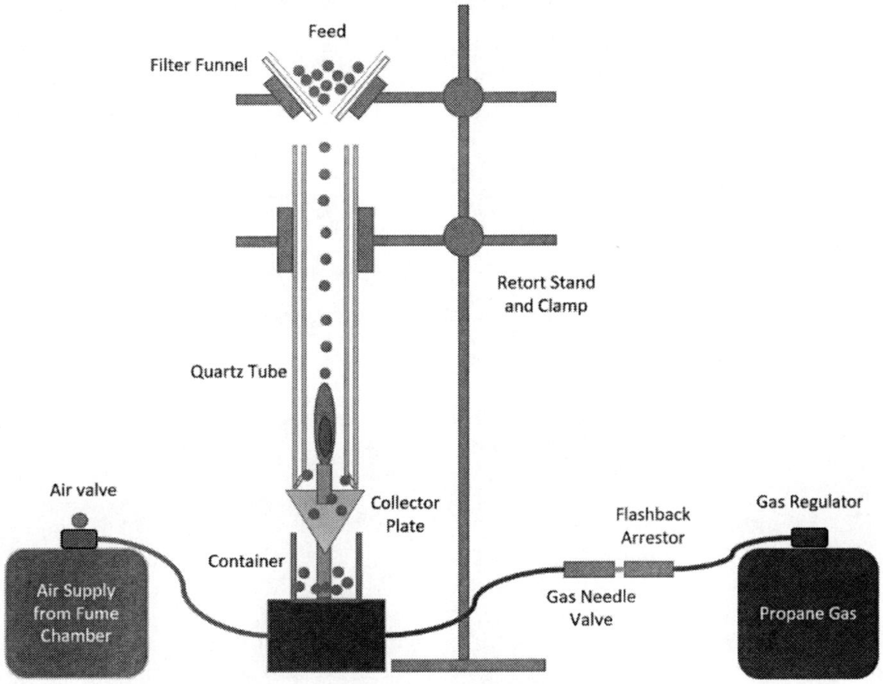

Figure 1. Vertical Thermal Flame Process [5]

2.3. Particle Characterization and Density Analysis

The powdered soda-lime glass and the respective thermally treated particles were characterized using a table-top Scanning Electron Microscope (SEM). SEM imaging was used to study the morphology of the powdered soda-lime glass and the respective thermally treated particles. A qualitative analysis is performed to determine if there is a density reduction of the pre-burned and burned soda-lime glass powder. Pre-burned soda-lime glass powder and burned soda-lime glass powder are placed in two different beakers (filled with distilled water). Floating particles in the beaker would show that there is density reduction of the burned particles.

3. Results and Discussion

Effects of thermal process on morphology of glass microsphere. A total of three samples between particle size selected are between 90 to 125μm are evaluated with single and second passes respectively. The effect of PVA as binding agent on the morphology of the glass microspheres was studied.

Figure 2 shows the morphology of pre-treated and treated particles. After the vertical thermal process, the morphology of some of the agglomerated soda-lime glass particles was observed to have transformation to spherical particle after the single pass of burning in the VTF. Figure 2 (a) shows the non-treated glass powder and Figure 2 (b) shows the morphology of agglomerated soda-lime glass powder with binder. Figure 2 (d) shows the second pass thermal treatment of the glass particles where it has better transformation to spherical particle as compared to the single pass thermal treatment of glass particles shown in Figure 2 (c). By undergoing two passes, it is found that the particles have spheroidized further. This is due to the longer overall contact time between the flame and particle, causing further spheroidization of the particles. This results in a more spherical particle as shown in Figure 2 (d).

Figure 2. Morphology of thermally treated glass powders: (a) powdered soda-lime glass (b) pre-treated soda-lime glass (c) formation of glass microspheres in single pass thermal treatment (d) formation of glass microspheres in two times pass thermal treatment

The formation of smooth spherical particle is dependent on the residence time of feed material in the flame produced. At the flame, the glass powder softens and spheroidizes. This is due to the effect of surface tension resulting in the formation of the spherical particle shape. As the temperature increase, the viscosity of the glass decreases. It is believed that when the glass is fed into the flame, the glass powder melts and gases are evolved forming the hollow cavity in the microsphere. As the glass powders exits the high temperature zone, the glass powders are cooled rapidly, conforming to the spherical particle shape.

Soda-lime glass have a relatively low melting point. The flame temperature supplied by the propane gas is sufficient to melt the soda-lime glass and decompose the gas in the glass particle resulting in the morphology change of the soda-lime glass from irregular surface to smooth spherical microspheres as shown in Figure 2 (a) to (d). The temperature produced by the propane gas ranges between 1000°C to 1500°C. The melting point of the soda-lime glass is at 1400°C. At this temperature, the glass melts and thus, forming the smooth spherical particles.

Similarly, Poirier *et.al* [6] produced glass microspheres by using oxy-acetylene flame that burns at a higher temperature and successfully produced smooth spherical particles. By comparison, oxy-acetylene gas burns at a higher temperature as compared to propane gas. Figure 2 shows production of glass microspheres by using the VTF set-up with propane gas which burns at a lower temperature than oxy-acetylene gas. This would suggest that glass microspheres can be produced using propane gas.

A comparison study is made with the production of hollow glass microsphere from amber glass frit by flame spraying method using oxy-acetylene gas. Dalai *et.al* incorporated urea as blowing agent in modification of the microsphere wall thickness and porosity [3]. Addition of urea as blowing agent showed there are changes the glass morphology. The morphology of the glass microsphere in Figure 1 (c) shows similar results to the hollow glass microsphere by Dalai *et al.* The successful production of glass microsphere as shown in Figure 1 (c) and Figure 1 (d) suggests that PVA can be used as a binder and blowing agent to aid in the development of the hollow cavity in the glass microsphere.

Qualitative analysis of pre-burned and burned particles on density reduction. Glass microspheres usually have density less than density of water [1]. Based on the literature reviews, thermally treated glass powders will have density less than 1000 kg/m^3 thus allowing it to float on water [7]. In contrast, pre-burned soda-lime glass powders with typical density of 2720 kg/m^3 [8] will sink in water. This would suggest that the glass microspheres will float on the surface of water. In Figure 3 (b), it is observed that the burned particles floats in a beaker of water as compared to the pre-burned particles in Figure 3 (a) where the pre-burned particles sinks at the bottom of the beaker. It is believed that there is a density reduction of the pre-burned particles. Therefore, it is believed that the VTF process have successfully formed hollow glass microsphere. The reduction in the density of the thermally treated glass particles depicts that the particles are hollow. These hollow particles can be used as lightweight fillers in syntactic foam or used as hydrogen gas storage in the pharmaceutical industry [3].

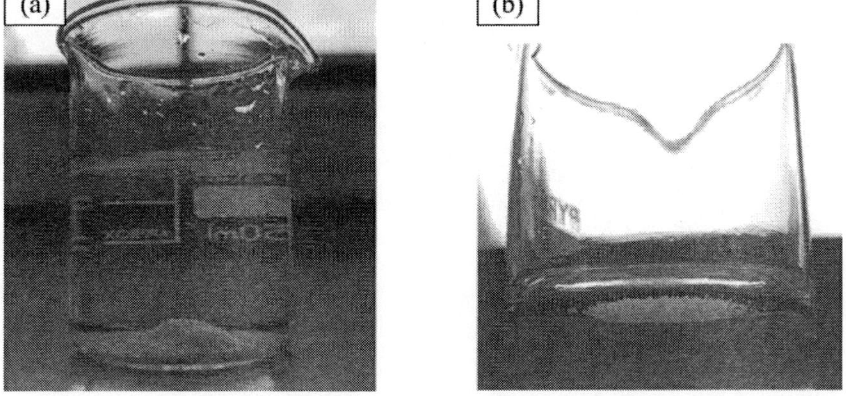

Figure 3. (a) Sink agglomerated soda-lime glass powder before undergoing the VTF process (b) floated burned particles after the VTF process

4. Conclusion

The morphology changed of the binded powdered soda-lime glass with PVA from an irregular surface to smooth spherical surface was observed from the SEM analysis. Hence, vertical thermal flame process is a feasible process to produce smooth spherical particle from powdered soda-lime glass. Other analytical test such as density measurement is recommended. A quantitative analysis would prove if there is a density reduction in the microspheres formed, proving if the microspheres formed are hollow or solid. Further improvements of the VTF process is suggested to increase the transformation of glass microspheres by using a vacuum system in the set-up or using a different source of heat (plasma or oxy-acetylene flame gun).

5. Acknowledgement

This work was supported by the collaboration between UCSI University Trust and Imperial College London

References

[1] Stuart M L 1993 *Handbooks of Composite Reinforcements* 3718 Cass Way Palo Alto, California 94306 VCH Publishers pp 248-50
[2] Watkins I G and Prado M 2015 **8** 1057-65
[3] Dalai S, Vijayalakshmi S, Shrivastava P, Sivam S, Sharma P 2014 **126** 65-70
[4] Ryzhenkovv A, Pogorelov S, Loginova N, Belvaeva E, Plestsheva A 2015 **9** 1913-52
[5] Soh W M, Tan J, Heng Jerry YY, Cheeseman C 2017 **880** 7-10
[6] Poirier T, Labrador N, Gamarra M, Enet N, Lira J 2005 **9** 509-20
[7] Budov V, Stetsenko V 1989 289-291
[8] Canakci H, Al-Kaki A, Celik F 2016 **161** 600-605

Titania with Alkaline Treated Graphitic Carbon Nitride (g-C_3N_4) to Improve Photocatalysis Properties

Wei Han Tan[1], Chen Hong Hak[1], Pichiah Saravanan[2], Kah Hon Leong[1], Lan Ching Sim[1]

[1]Department of Environmental Engineering, Faculty of Engineering and Green Technology, Universiti Tunku Abdul Rahman, Kampar, Perak, Malaysia.
[2]Department of Environmental Science and Engineering, Indian Institute of Technology, Dhanbad, India.
ch.hak117@gmail.com

Abstract. The graphitic carbon nitride (g-C_3N_4) was treated via alkaline hydrothermal treatment to change the porous structure of g-C_3N_4 to "tube-like" structure. The developed alkaline treated g-C_3N_4 (A-g-C_3N_4) was combined with P25 Degussa TiO_2 for the competent removal of methylene blue (MB). The morphological changes increased the crystallite size and pore size of g-C_3N_4, causing a decrease in the specific surface area of A-g-C_3N_4. The pure g-C_3N_4 demonstrated the best degradation efficiency among the all samples due to its high specific surface area, low band gap energy and small pore size. The combination of both A-g-C_3N_4 and g-C_3N_4 with TiO_2 did not exert significant effect on the degradation efficiency of MB owing to the low specific surface area, high band gap and large pore size. Thus concluding the degradation efficiency of organic dye is attributed predominantly to the factors of energy band gap, specific surface area and pore sizes.

1. Introduction

TiO_2 is commonly used as photocatalyst for reductive and oxidative reactions on its surface among the semiconductors. To boost the photocatalytic efficiency of TiO_2, the mass transfer limitation has to be minimized since photocatalytic reaction mainly occurs on the surface of TiO_2. The adsorption of organic pollutants on TiO_2 surface is restricted by its poor affinity towards the pollutants. Therefore, how to concentrate the target pollutants around the TiO_2 nanoparticles to enhance photocatalytic efficiency requires consideration. The TiO_2 nanoparticles may undergo aggregation due to the instability of the nanosized particle during the photocatalytic degradation process which may hamper the light incidence on the active centers and consequently reduce its catalytic activity [1,2]. TiO_2 has been highly employed for commercial purposes, due to its better photoactivity and stability under ultraviolet (UV) radiation i.e., $\lambda < 400$ nm. Though it has above mentioned advantages the major limitations are: a) It is active only under UV light and can only absorb about 5% of sunlight in the UV region [3] and b) high recombination rate of photoinduced charge carriers cause TiO_2 unable to apply efficiently [4]. Graphitic carbon nitride material (g-C_3N_4) has recently emerged as a novel photocatalyst attributed to its visible-light driven band gap (2.69 eV), high thermal and chemical stability, and facile synthesis method. However, the surface area of g-C_3N_4 is relatively small (14.3 m^2/g) [5]. It is considered that the small specific surface area of g-C_3N_4 could contribute to the slow photocatalytic reaction rate [6]. The specific surface area of g-C_3N_4 need to be increased to make g-C_3N_4 become a promising and versatile semiconductor for sunlight harvesting [7,8].

2. Preparation of g-C_3N_4 and A-g-C_3N_4

20 g of urea (R & M Marketing Essex, U.K) was dried in an oven at 80 °C for 1 day and then put in a muffle for 3 h at 500 °C. The yellowish product was washed with nitric acid (0.1 M) several times and distilled water to remove any residual alkaline species. The g-C_3N_4 and 90 cm^3 of NaOH (0.10 mol dm^{-3}) solutions were mixed in Teflon-lined autoclave and subjected to hydrothermal treatment at 150 °C for 18 h. The solid product was then dried at 80 °C for 24 h and the as-obtained sample was denoted as A-g-C_3N_4.

2.1. Synthesis of g-C_3N_4/A-g-C_3N_4 hybridized TiO_2

0.15 g prepared g-C_3N_4/A-g-C_3N_4 sheet was well dispersed in distilled water ultrasonically. 4.8 g of freshly prepared P25 Degussa TiO_2 was then added into the solution and subjected to 70 °C for 1 h. The resulting suspension was then centrifuged and washed repeatedly with distilled water for few times and dried overnight at 60 °C. The sample obtained by synthesized was denoted as g-C_3N_4/TiO_2 and A-g-C_3N_4/TiO_2, respectively.

2.2. Characterization

The powder X-ray diffraction (XRD, PANalytical-Empyrean) patterns were acquired with Cu Kα radiation at a scanning speed of $0.02s^{-1}$. High resolution transmission electron microscope (HRTEM, JEM-2100F, Jeol) images were obtained at 200 kV. The EDX analysis of the A-g-C_3N_4/TiO_2 was observed on a field emission scanning electron microscope (FESEM JEOL 6701-F). The Brunauer–Emmett–Teller specific surface area and pore volume of samples was determined at liquid nitrogen temperature (77 K) based on nitrogen adsorption-desorption isotherms with TriStar II 3020 (Micrometrics®, USA).

2.3. Photocatalytic degradation of organic pollutants

0.02 g of photocatalysts were immersed in a glass beaker containing 250 mL aqueous solutions for MB (5 ppm). Prior to photodegradation and to establish an adsorption-desorption equilibrium, the solutions were magnetically stirred in a dark for 1 h. A 500 W tungsten-halogen lamp was used as visible light source with filter glass to filter UV light ($\lambda < 420$nm). The dye samples were collected at regular interval, then analyzed for residual MB concentration with visible spectrophotometer (DR 6000 ™ UV-Vis spectrophotometer (Hach)) to measure the intensity of light passing through MB (λ_{max}= 664 nm).

3. HRTEM and EDX

In figure 1, it is clearly seen that TiO_2 nanoparticles were well distributed on the surface of g-C_3N_4. The surface of A-g-C_3N_4 was aggregated with nanoclusters of TiO_2 due to the nanometer effect [9]. The EDX spectrum for elemental compositions of the g-C_3N_4/TiO_2 is illustrated in figure 1b, confirming the prepared sample was a constituent of C, Ti, O and N.

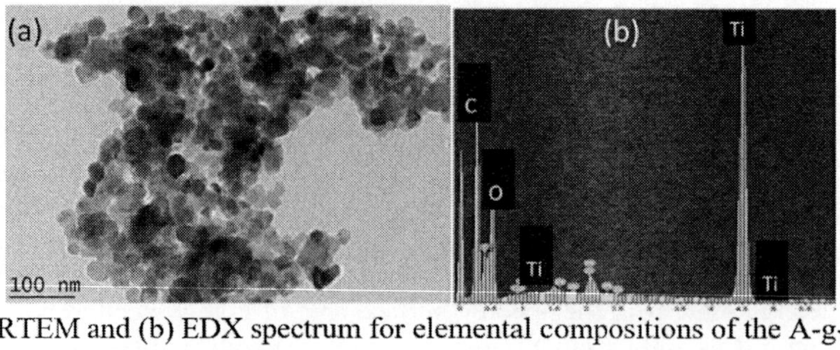

Figure 1. (a) HRTEM and (b) EDX spectrum for elemental compositions of the A-g-C_3N_4/TiO_2

4. Crystallite size and BET surface

The average crystallite sizes of TiO_2 anatase and g-C_3N_4 were calculated using Scherrer equation:

$$D = \frac{K\lambda}{\beta \cos\theta} \tag{1}$$

where K is the shape factor taken as 0.89 for calculations, λ is the wavelength of X-ray (0.154 nm), β is the full width half maximum (FWHM) for the 2θ peak, and θ is the diffraction angle. The crystallite size of TiO_2 anatase and g-C_3N_4 is listed in table 1. After incorporating both g-C_3N_4 and A-g-C_3N_4 in binary composites, it is observed that the crystallite size of TiO_2 decreased while the crystallite size of g-C_3N_4 increased. The size of TiO_2 was inhibited by the g-C_3N_4 nanosheet since TiO_2 nucleated and grew on the g-C_3N_4 surface [8]. The alkaline hydrothermal treatment of g-C_3N_4 significantly increased the crystallize size of A-g-C_3N_4 to 43.17 nm and thus, a decreasing specific surface area could occur in this sample.

Table 1: Average crystallite size for P25, A-g-C_3N_4/TiO_2, g-C_3N_4, A-g-C_3N_4 and g-C_3N_4/TiO_2

Sample	Crystallite size(nm)[a]	Crystallite size(nm)[b]
P25	32.35	-
A-g-C_3N_4/TiO_2	24.7	43.17
g-C_3N_4	-	5.27
A-g-C_3N_4	-	18.68
g-C_3N_4/TiO_2	25.26	35.04

a: TiO_2
b: g-C_3N_4

According to the International Union of Pure and Applied Chemistry (IUPAC) classification, all of the samples in figure 2 demonstrated adsorption-desorption isotherms of type III which indicating the existence of mesopores. Characteristic of type III was the heat of adsorption less than the adsorbate heat of liquification, adsorption and the interaction with the adsorbent surface is lower than the adsorbate interaction with an adsorbed layer [10]. The shape of the hysteresis loop is Type H2 when the low P/P_0 range from 0.5 to 0.8, indicating the existence of ink-bottle pores with broad bodies and narrow necks [5].

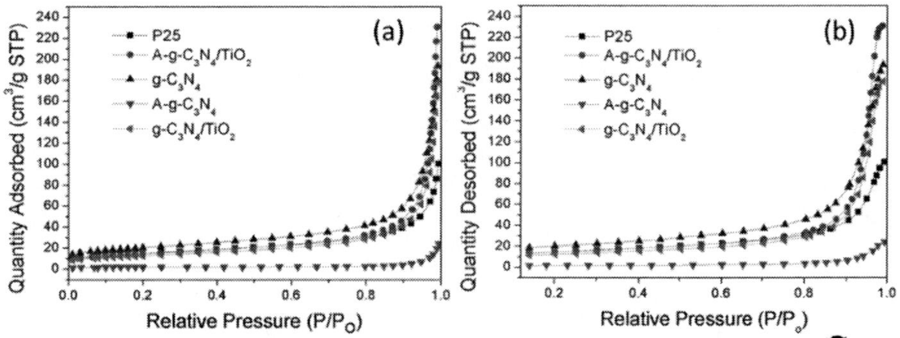

Figure 2. (a) Adsorption and (b) desorption isotherms of P25, A-g-C_3N_4/TiO_2, g-C_3N_4, A-g-C_3N_4 and g-C_3N_4/TiO_2 measured at 77 K

5. Photocatalytic Experiment

Figure 3 depicts visible-light-induced photocatalytic activity with the aid of prepared photocatalysts. The photocatalytic performance followed an order of g-C_3N_4 (62.67%) > P25 (49.47%) > A-g-C_3N_4 (47.33%) > A-g-C_3N_4/TiO_2 (46.87) > g-C_3N_4/TiO_2 (45.79%). The g-C_3N_4 showed higher photodegradation efficiency which was 62.67% compared to that of other samples, while g-C_3N_4/TiO_2 exhibited the lowest photodegradation efficiency which was 45.79% (table 2). In addition, the large specific surface area of g-C_3N_4 (71.8 m^2g^{-1}) could provide more active adsorption sites for the reactant, and thus leading to superior degradation efficiency. The photodegradation efficiency of g-C_3N_4/TiO_2

was slightly lower than that of A-g-C_3N_4/TiO_2 and A-g-C_3N_4 due to its specific area was smaller than A-g-C_3N_4/TiO_2 and its band gap was higher than A-g-C_3N_4.
Due to the large band gap and fast recombination rate of photo-generated electron-hole pairs, the other sample showed lower photodegradation efficiency than g-C_3N_4 [11]. The narrowest band gap in pure A-g-C_3N_4 did not lead to the enhancement of photocatalytic performance due to its small specific surface area, which was 6.3 $m^2 g^{-1}$. The photocatalytic activity performance was related to the specific area of the photocatalyst, the higher the specific surface area of the photocatalyst, the higher the photocatalytic activity [12]. The samples of A-g-C_3N_4/TiO_2, A-g-C_3N_4 and g-C_3N_4/TiO_2 showed an outstanding visible light harvesting properties, but their large pore sizes ranging from 230 to 270 nm and low specific surface area large contributed to the poorer photocatalytic activity compared to that of P25 TiO_2. The observed degradation data were fitted to the simple kinetic model and tabulated in table 2. The first-order reaction kinetics is expressed by equation:

$$\ln\left(\frac{C}{C_o}\right) = -kt \qquad (2)$$

where k is the first-order reaction constant, C_0 and C are the MB concentrations in solution at times 0 and t, respectively.

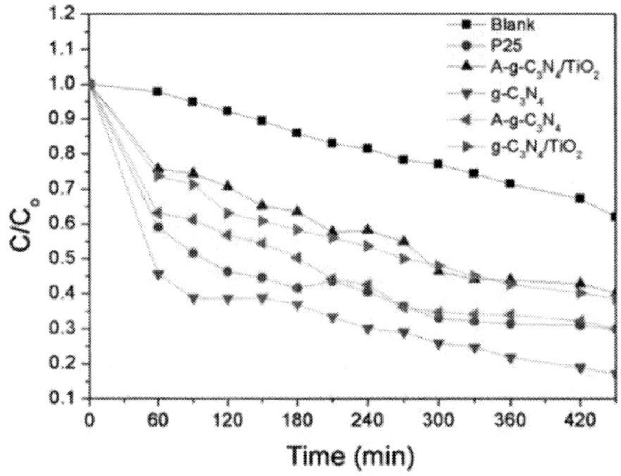

Figure 3. Photocatalytic degradation of MB as a function of reaction time

Table 2: Degradation rate of MB by P25, A-g-C_3N_4/TiO_2, g-C_3N_4, A-g-C_3N_4 and g-C_3N_4/TiO_2

Sample	Degradation Rate[a] (%)	K^b (min^{-1})
Blank(control)	32.78	0.001
P25	49.47	0.002
A-g-C_3N_4/TiO_2	46.87	0.002
g-C_3N_4	62.67	0.003
A-g-C_3N_4	47.33	0.002
g-C_3N_4/TiO_2	45.79	0.002

[a] after reaction for 6 h
[b] apparent rate constant deduced from linear fitting of ln(C/C_0) versus reaction time

6. Conclusion

In summary, the binary composites namely A-g-C_3N_4/TiO_2 and g-C_3N_4/TiO_2 was successfully synthesized *via* a simple and facile method. The prepared binary composites displayed non appreciable changes in the photocatalytic performance compared to that of pure g-C_3N_4. This is mainly due to the decreasing specific surface area in binary nanocomposites when the titania agglomerated on the surface of both g-C_3N_4 and A-g-C_3N_4. The overetching effect caused by the non-optimized conditions of alkaline hydrothermal treatment also contributed to a trivial photocatalytic activity of A-g-C_3N_4.

Hence, the specific surface area and crystallite size played a crucial role in boosting the photocatalytic performance of photocatalysts.

References

[1] Gao B, Yap P S, Lim T M and Lim T T 2011 Adsorption-photocatalytic degradation of acid red 88 by supported TiO_2: effect of activated carbon support and aqueous anions *Chem. Eng. J.* **171(3)** 1098-1107.

[2] Mallakpour S and Nikkhoo E 2014 Surface modification of nano-TiO_2 with trimellitylimido-amino acid-based diacids for preventing aggregation of nanoparticles *Adv. Powder Technol.* **25(1)** 348-353.

[3] Dong H, Zeng G, Tang L, Fan C, Zhang C, He X and He Y 2015 An overview on limitations of TiO_2-based particles for photocatalytic degradation of organic pollutants and the corresponding countermeasures *Water Res* **79** 128-146.

[4] Bagwasi S, Niu Y, Nasir M, Tian B and Zhang J 2013 The study of visible light active bismuth modified nitrogen doped titanium dioxide photocatlysts: role of bismuth *Appl. Surf. Sci.* **264** 139-147.

[5] Hao R, Wang G, Tang H, Sun L, Xu C and Han D 2016 Template-free preparation of macro/mesoporous g-C_3N_4/TiO_2 heterojunction photocatalysts with enhanced visible light photocatalytic activity *Appl. Catal. B: Environ.* **187** 47-58.

[6] Sano T, Tsutsui S, Koike K, Hirakawa T, Teramoto Y, Negishi N and Takeuchi K 2013 Activation of graphitic carbon nitride (g-C_3N_4) by alkaline hydrothermal treatment for photocatalytic NO oxidation in gas phase *J. Mater. Chem.* A **1(21)** 6489-6496.

[7] Chen X, Wei J, Hou R, Liang Y, Xie Z, Zhu Y, Zhang X and Wang H 2016 Growth of g-C_3N_4 on mesoporous TiO_2 spheres with high photocatalytic activity under visible light irradiation *Appl. Catal. B: Environ.* **188** 342-350.

[8] Tong Z, Yang D, Xiao T, Tian Y and Jiang Z 2015 Biomimetic fabrication of g-C_3N_4/TiO_2 nanosheets with enhanced photocatalytic activity toward organic pollutant degradation *Chem. Eng. J.* **260** 117-125.

[9] Li C, Sun Z, Xue Y, Yao G and Zheng S 2016 A facile synthesis of g-C_3N_4/TiO_2 hybrid photocatalysts by sol–gel method and its enhanced photodegradation towards methylene blue under visible light *Adv. Powder Technol.* **27(2)** 330-337.

[10] Leddy N 2012 Surface area and porosity Available at: http://cma.tcd.ie/misc/Surface_area_and_porosity.pdf [Accessed 17 August 2016]

[11] Faisal M, Ismail A A, Harraz F A, Al-Sayari S A, El-Toni A M and Al-Assiri M S 2016 Synthesis of highly dispersed silver doped g-C_3N_4 nanocomposites with enhanced visible-light photocatalytic activity *Mater. Design* **98** 223-230.

[12] Li J, Liu Y, Li H and Chen C 2016 Fabrication of g-C_3N_4/TiO_2 composite photocatalyst with extended absorption wavelength range and enhanced photocatalytic performance *J. Photoch. Photobio. A* **317** 151-160.

IOP Conf. Series: Materials Science and Engineering **205** (2017) 012024 doi:10.1088/1757-899X/205/1/012024

Thermal Treatment Temperature and Time Dependence of Contact Angle of Water on Fluorinated Polystyrene as Hydrophobic Film Coating

M S Tolentino[1], J F Carpena[1], R M Javier[1] and R R Aquino[1]

[1]School of Chemical Engineering and Chemistry, Mapua Institute of Technology, Manila, Philippines, 1002
rraquino@mapua.edu.ph

Abstract. The study focuses on the synthesis of fluorinated polystyrene (F-PS) as a hydrophobic film coating and on the investigation of the consequent effects of thermal treatment time and treatment temperature on its contact angle. The fluorinated polystyrene is synthesized via Friedel-Crafts acylation of the benzoic rings by electrophilic substitution of trifluoroacetic anhydride in the presence of $AlCl_3$ in an environment of dichloromethane as the aprotic solvent with a reaction temperature of 36°C. The reaction yielded fluorinated polystyrene characterized by a solid brown substance. FT-IR analysis of the substance had shown band peaks at 1150 cm^{-1}, a wavenumber indicating the presence of fluorine in the synthesized material, which lowers its surface energy. SEM images of the F-PS show nucleation sites giving rise to hierarchical structures on the surface of the material due to the action of fluorination. The contact angle of fluorinated polystyrene, upon application of thermal treatment, increased to as much as 29.26% when compared to the unmodified polystyrene manifesting a preferentially more hydrophobic behavior. It was also found that the contact angle increases linearly with treatment temperature while statistical analysis shows that thermal treatment time has no significant effect on the hydrophobicity of the F- PS.

1. Introduction

Due to the numerous impending industrial applications of hydrophobic and oleophobic characteristics of different surfaces, researchers became interested with the factors affecting the hydrophobicity of such surfaces [1-4].

The main purpose of this study is to fabricate a hydrophobic film coating from fluorinated polystyrene (F-PS) to improve its hydrophobic behavior upon the application of heat treatment. Specifically, this undertaking seeks to achieve the following: (1) to fluorinate polystyrene in order to increase its hydrophobicity (2) to characterize the surface morphology of the fluorinated polystyrene film through the use of scanning electron microscopy (SEM); (3) to characterize the surface chemical composition of the fluorinated polystyrene film using Fourier Transform Infrared Spectroscopy (FT-IR); and (4) to determine the static contact angle of the fluorinated polystyrene film at varying temperature conditions and thermal treatment times.

2. Methodology

Fluorination of polystyrene (PS) was carried out through a synthetic route involving different reactions on the PS aromatic ring. This step was based on the mechanism proposed by Donchak and Harhay [5].

Several modifications on parameters and solvents used in the said reaction mechanism were employed in the present study.
Through electrophilic substitution under the Friedel-Crafts condition, the benzoic rings of polystyrene were acylated by trifluoroacetic anhydride (TFA) in an environment of dichloromethane (DCM) as the aprotic anhydrous solvent with aluminum chloride ($AlCl_3$) as the Lewis acid catalyst. The sequence of reactions started with the loading of 3.33 g of PS pellets in a round bottom flask, which were then dissolved under heat (36°C) with 100 ml DCM. The flask was immersed in a water bath to avoid direct heating. After complete dissolution, it was allowed to be cooled to 20°C followed by the addition of TFA (13.32 g) with 2:1 molar ratio to PS. $AlCl_3$ catalysts (8.43 g) were then added in small quantities. Maintaining the same temperature, the system was mixed for one hour to ensure efficient contact of the polystyrene rings with the catalyst. It was then subjected under heat (36°C) where it was mixed for two hours. The resulting suspension had a characteristic brown color after the mixing period. It was cooled to room temperature and then 15 ml dimethylformamide (DMFA) was added to dissolve the suspension. The suspension was then filtered using a vacuum pump and the resulting residue was repeatedly washed with 10 ml DCM. The resulting filtrate was added to excess amounts of ether (preferably 250 ml) and was refrigerated to recrystallize the fluorinated polystyrene (F-PS) characterized by a brown solid substance. After recrystallization in ether, the solid was filtered from the ether, air-dried under ambient conditions, and oven-dried at 80 °C for 5 minutes to vaporize all adhering solution yielding 0.75 g of brown solid F-PS. The preceding reactions are shown by the equation:

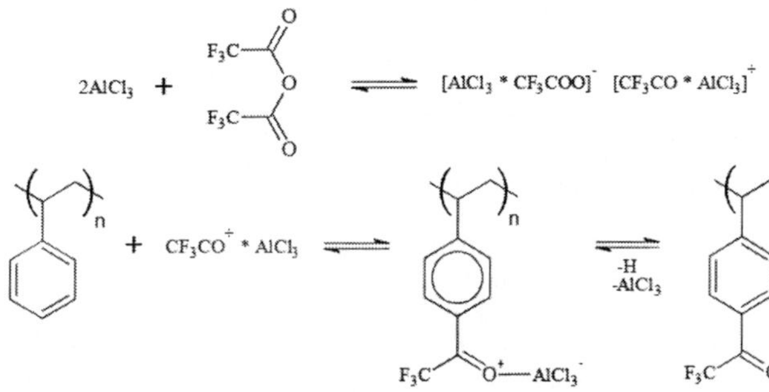

Figure 1. The Friedel-Krafts acylation for the fluorination of polystyrene.

To determine if the fluorination of polystyrene is successful, the sample was subjected to Fourier Transform Infrared Spectroscopy. To analyze the surface morphology, Scanning Electron Microscopy was employed. A contact angle goniometer was used to measure the angle between the surface and a tangent line of the droplet connected to the interface.

3. Results and Discussion

3.1. Fourier Transform Infrared Spectroscopy

Small amounts of the final product were tested using FT-IR spectroscopy. The Attenuated Total Reflectance (ATR) sampling technique was used in this step for the solid sample. Figure 2 shows a portion of the FT-IR results of the fluorinated polystyrene (F-PS). Band peaks at 1150 /cm and 940 /cm wavenumber indicates the presence of fluorine in the synthesized material.

IOP Conf. Series: Materials Science and Engineering **205** (2017) 012024 doi:10.1088/1757-899X/205/1/012024

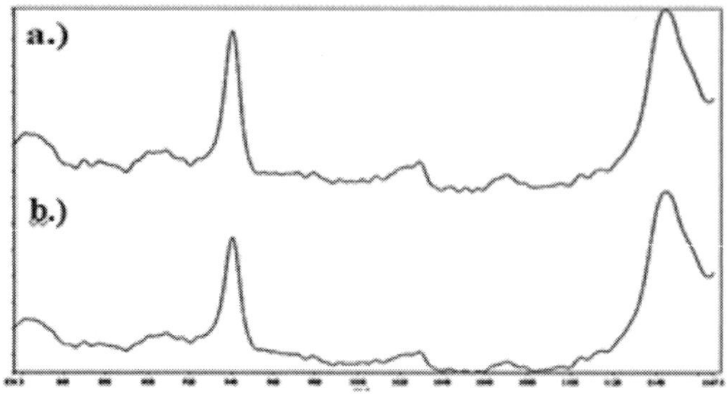

Figure 2. FT-IR spectra for (a) the synthesized F-PS and (b) the unmodified PS.

3.2. Scanning Electron Microscopy

The surface morphology of the fluorinated polystyrene (F-PS) was analyzed through SEM, and was compared to the surface morphology of the polystyrene (PS). Images of the surface of fluorinated and unmodified polystyrene were taken using an SNE 3200M model scanning electron microscope at a magnification power of 100x, 1000x and 3000x. It can be seen that rough structures are present on the surface of the F-PS, while a smooth surface morphology can be seen for the PS. Comparing the images of F-PS with that of PS, it can be deduced that the presence of fluorine makes the F-PS surface structure rough.

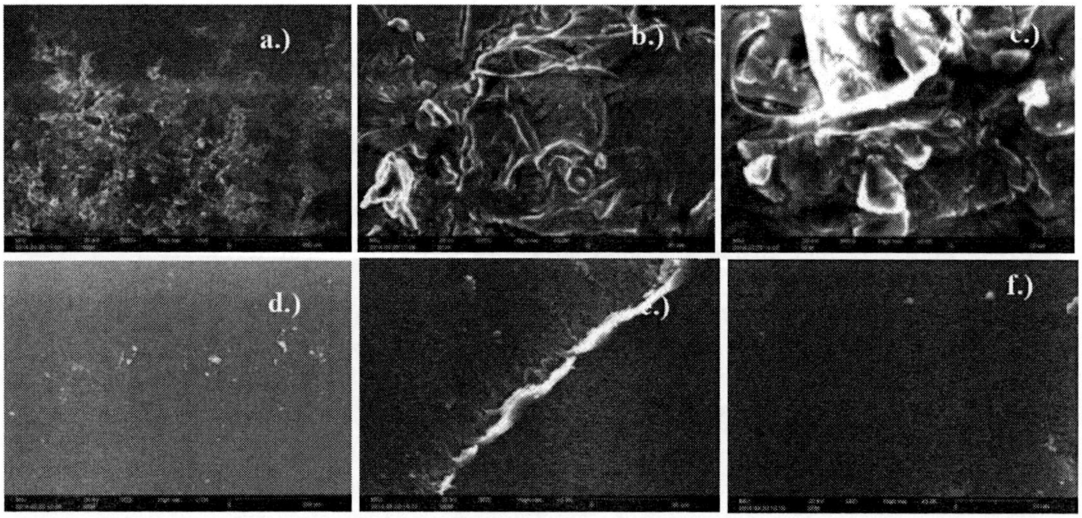

Figure 3. SEM images of F-PS (a,b,c) and PS (d,e,f) at 100x, 1000x, and 3000x magnifications respectively.

3.3. Effect of Thermal Treatment Time and Temperature on the Hydrophobicity of the Fluorinated Polystyrene Surface Coating

In order to quantify the effects of thermal treatment time and temperature in the hydrophobicity of the fluorinated polystyrene surface coating, its static contact angle were measured right after the treatment procedure. For comparative purposes, unmodified commercial-grade polystyrene material from Chemrez® was also subjected to the same treatment and its subsequent contact angle was measured as well.

Figure 4 shows that the contact angle of the fluorinated polystyrene (F-PS) increases with the temperature at all treatment times.

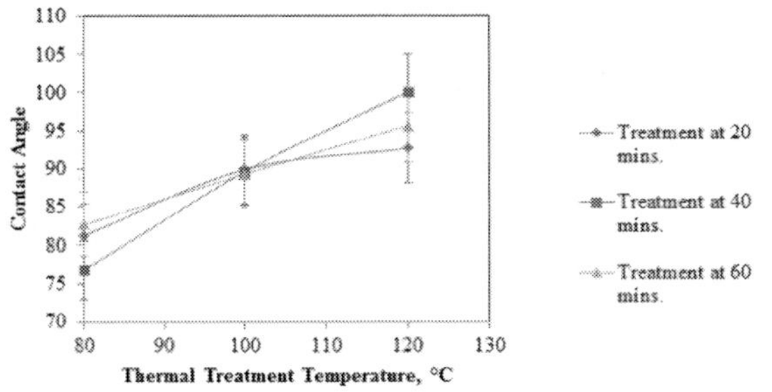

Figure 4. Treatment temperature dependence of contact angle of water on fluorinated PS.

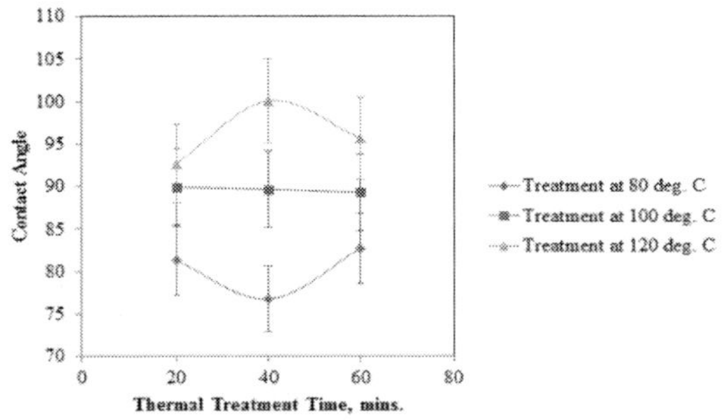

Figure 5. Treatment time dependence of contact angle of water on fluorinated PS

When the relationship of contact angle and treatment time is analyzed (Figure 5), it is apparent that there is no significant increase in contact angle when the treatment temperature is held constant. Nevertheless, the contact angles of the surface treated at 100°C and 120°C shows more hydrophobic character.

3.4. Effect of Thermal Treatment Time and Temperature on the Hydrophobicity of the Unmodified Polystyrene Surface Coating

The contact angle measurements at different treatment temperature for each treatment time of the unmodified polystyrene (PS) material are shown on Figure 6. As the temperature is increased from 80°C to 120°C, the contact angle also increases within the range of 70° to 80°. Moreover, with regards to the relationship of contact angle and treatment time for PS, the same observations, with those of F-PS, transpired. Based on Figure 7, no significant increase in contact angle was observed at constant treatment temperature.

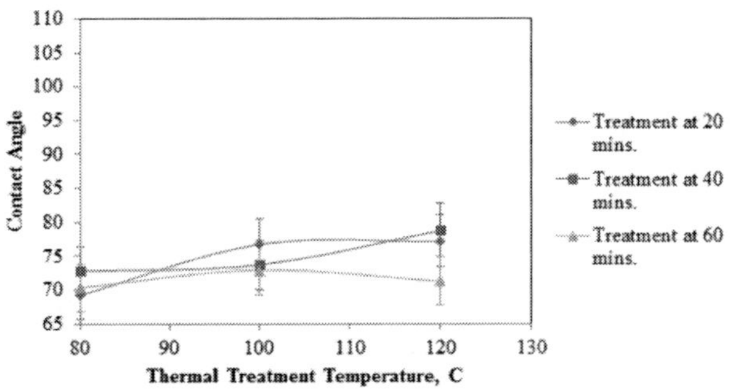

Figure 6. Treatment temperature dependence of contact angle of water on PS.

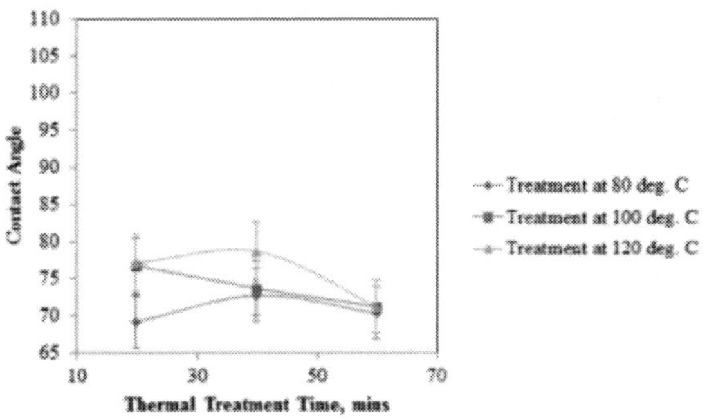

Figure 7. Treatment time dependence of contact angle of water on PS.

Upon comparison of the average contact angles of the fluorinated polystyrene and unmodified polystyrene, it is apparent that the F-PS has a more hydrophobic character. Furthermore, the contact angles are more significantly increased as the temperature is increased as well. For the 20-minute treatment time, the contact angle of the starting material (PS) is increased to as much as 18.24% at 120°C treatment after fluorination.

On the other hand, the contact angle is increased further to 23.86% for the fluorinated polystyrene, when treated at 40 minutes at 120°C. Likewise, at 60 minutes treatment time at 120°C, the contact angle of the F-PS and PS has a difference of 29.26%, representing an increase in hydrophobicity of the material upon fluorination (Figure 8).

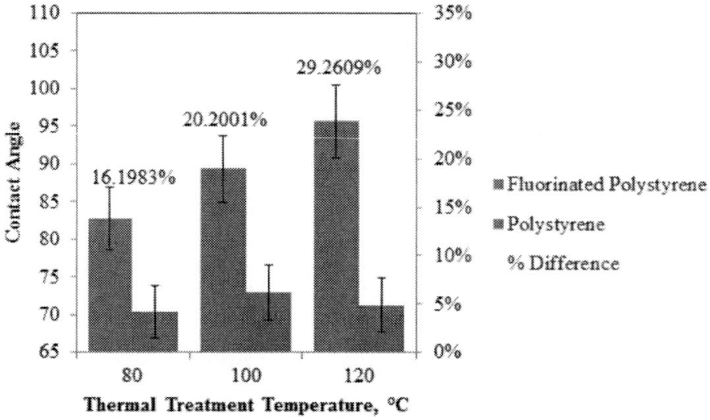

Figure 8. Comparison of contact angle of water on PS vs. Fluorinated PS at 60 minutes treatment.

4. Conclusion

Based on the results and analysis of this study, it is concluded that the results of Fourier Transform Infrared Spectrometer show that fluorine atoms are present in the synthesized material as indicated by the band peak at 1150 cm^{-1} wavenumber of the spectra. The images obtained through the scanning electron microscope show the apparent differences in the surface of the unmodified polystyrene and F-PS. From the said images, it can be inferred that the surface of the fluorinated polystyrene shows a rougher surface due to the nucleation sites upon fluorination. On the other hand, the surface of the unmodified PS shows a smooth texture revealing its high wetting behavior.

Thermal treatment temperature is concluded to be a strong factor for the increase in contact angle of the fluorinated polystyrene film coating. It can then be said that no matter what the treatment time is, there is no significant effect in the resulting contact angle of the F-PS since all of the contact angle readings obtained lie relatively within the 5% threshold of each other when measured at different thermal treatment times. Inducing the fluorinated polystyrene coating to a high temperature environment also increases the corresponding contact angle of water on its surface. Based on the findings of this research, the hydrophobicity of the surface (in terms of its contact angle) can be increased to as much as 29.26% when treated at 120°C. The contact angle also shows a direct variation with temperature.

References

[1] Godeau G, Amigoni S, Darmanin T and Guittard F 2016 Post-functionalization of plasma treated polycarbonate substrates: An efficient way to hydrophobic, oleophobic plastics *Applied Surface Science* **387** 28-35

[2] Joki-Korpela F, Karvinen J, Paivanranta B, Partamen A, Suvanto M, Kuittinen M and Pakkanen T 2014 Hydrophobic and oleophobic anti-reflective polyacrylate coatings *Microelectronic Engineering* **114** 38-46

[3] Li H, Yu S, Han X, Liu E and Zhao Y 2015 Fabrication of superhydrophobic and oleophobic surface on zinc substrate by a simple method *Colloids and Surfaces A: Physicochemical and Engineering Aspects*, **469** 271-278

[4] Parale V , Mahadik D , Phadtare V , Pisal A , Park H and Wategaonkar S 2016 Dip coated superhydrophobic and anticorrosive silica coatings *International Journal of Materials Science and Engineering,* **4** 60-68

[5] Donchak V and Harhay K 2008 Synthesis of Fluorinated Polystyrene *Chemistry and Chemical Technology* 11-14

IOP Conf. Series: Materials Science and Engineering **205** (2017) 012025 doi:10.1088/1757-899X/205/1/012025

Effect of Cooling Rates on Shape and Crystal Size Distributions of Mefenamic Acid Polymorph in Ethyl Acetate

S. K. Abdul Mudalip[1*], F. Adam1, J. Parveen[2], M. R. Abu Bakar[3], N. Amran1, S. Z. Sulaiman[1], R. Che Man[1], Z. I. Mohd Arshad[1], S. Md. Shaarani[1]

1 Faculty of Chemical Engineering & Natural Resources, Universiti Malaysia Pahang, 26300 Gambang, Pahang.
2 Department of Biotechnology Engineering, Kulliyyah of Engineering, International Islamic University Malaysia, 50728 Kuala Lumpur, Malaysia.
3 Department of Pharmaceutical Technology, Kulliyyah of Pharmacy, International Islamic University Malaysia, Bandar Indera Mahkota, 25200 Kuantan, Pahang, Malaysia.
*kholijah@ump.edu.my

Abstract. This study investigate the effect of cooling rates on mefenamic acid crystallisation in ethyl acetate. The cooling rate was varied from 0.2 to 5 °C/min. The in-line conductivity system and turbidity system were employed to detect the onset of the crystallization process. The crystals produced were analysed using optical microscopy and Fourier transform infrared spectroscopy (FTIR). It was found that the crystals produced at different cooling rates were needle-like and exhibit polymorphic form type I. However, the aspect ratio and crystal size distributions were varied with the increased of cooling rate. A high crystals aspect ratio and narrower CSD (100–900 μm) was obtained at cooling rate of 0.5 °C/min. Thus, can be suggested as the most suitable cooling rate for crystallization of mefenamic acid in ethyl acetate.

1. Introduction

Crystallization is a process that typically used to obtain solid material in purified form. It was reported that about 80% of the active pharmaceutical ingredients (APIs) are produced using at least one crystallization step [1]. The main challenge in the crystallization process is to produce crystals with desired properties such as a polymorphic form, morphology and crystal size distribution [2]. Polymorphs of a API can show different physicochemical properties particularly the melting point, density, morphology, stability, bioavailability, and processability during the manufacturing process [3]. The shape of crystals and the CSD determine the quality of the final product and affect the downstream operations such as filtration and drying. For a needle-like shape crystals, the aspect ratio should be equal or greater than one to avoid any disappearance in the solution [4]. This is because crystals with high aspect ratio are susceptible to breakage during the filtration process [5]. Various process variables were reported to show the significant effect on the crystal properties during the cooling crystallization process. In order to get a desired crystal shape and CSD with good reproducibility in every batch, the operating conditions that affect these properties during the crystallization process must be controlled.

Mefenamic acid [2-(2, 3-dimethylphenyl)amino benzoic acid] is a non-steroidal anti-inflammatory and analgesic agent that widely used for management of pain during menstrual period. This API is reported as one of the active pharmaceutical ingredients (APIs) that exhibit polymorphisms and can either exist in Form I, Form II or Form III [6-8]. The mefenamic acid Form I, Form II and Form III are enantiotropic related, where Form I is relatively more stable than Form II and Form III at ambient condition [9]. In this work, the effect of cooling rates during batch crystallization process on the crystal shape factor namely aspect ratio and CSD mefenamic acid were investigated. In-line analysers; conductivity and turbidity probes, were used during crystallization process. Comprehensive characterization of crystals produced were performed using infrared spectroscopy (FTIR) and optical microscopy.

2. Methodology

2.1. Chemicals and Materials

The mefenamic acid powder (98 wt% purity) was obtained from Baoji Tianxin Pharmaceutical Co. Ltd., China. The ethyl acetate (99.5 wt% purity) was obtained from Permula Chemical (M) Sdn. Bhd., Pahang, Malaysia. The mefenamic acid and ethyl acetate were used without further purification.

2.2. Methodology

The crystallization experiments were performed in a 500 mL crystallization vessel. A saturation concentration was prepared in the crystallization vessel by dissolving about 1.81 g of mefenamic acid in 100 g of ethyl acetate. The solution was heated at 10 °C above the saturation temperature for 30 min. After complete dissolution of the solute is achieved, the solution was cooled to the final cooling temperature between 10 °C to 30 °C using a cooling rate of 0.5 °C/min. The crystallization process was stopped 30 min after the nucleation commences. A conductivity and turbidity system were employed for in-line detection of polymorph nucleation event during the solution crystallization process. The onset of the nucleation events is indicated either by the sudden decrease of the conductivity values or the sudden increase in the turbidity values. The obtained crystals were filtered and dried in the oven at 50 °C until constant dried weight was achieved. The dried crystals were stored in glass vials until further analysis. The same procedures were repeated using cooling rate of 0.2, 1.0, 2.5, and 5.0 °C/min.

2.3. Characterizations

The images of the crystals were captured using Leica microscope DM750 with a total magnification of 200x4x/0.10 and processed using Leica Application Suite Software version 3.6. For determination of the length and width of the crystals, image analysis toolbox available in MATLAB was used. Random measurement of 100 crystals were performed to determine the aspect ratio and crystal size distribution (CSD) (Liu et al., 2013). The FTIR spectra of crystals produced were acquired using a Perkin Elmer's ATR-FTIR Spectrometer (Frontier) with a wavenumber range of 500 to 4000 cm^{-1}. The analysis was performed using OMNIC software with an average of 16 scans [6].

3. Results & Discussion

Figure 1 (a) illustrates the partial FTIR spectrum of mefenamic acid crystals that crystallized using ethyl acetate at different cooling rates. As seen, the spectrum shows the presence of O-H and N-H bonds at wavelength of 2986 and near 3313 cm^{-1} and consistent with the IR adsorption spectra of mefenamic acid Form I reported by previous work [6]. The N-H stretching band occurs between 3300 and 3350 cm^{-1}, is an important spectral point that can be used to distinguish between Form I and Form II of mefenamic acid. Specifically, the N-H stretching frequency, which occurs at 3311 to 3313 and 3346 to 3350 cm^{-1}, show the presence of Form I and Form II, respectively. The N-H stretching at these wavelengths is observed due to the formation of internal hydrogen bonding between the amino group and the carbonyl group.

Figure 1 (b) and Figure 2 illustrate the change of crystals aspect ratio and CSD, respectively with different linear cooling rates for solution concentration of 1.81 g/ 100 g ethyl acetate and crystallization time of 30 min. It can be observed that the highest cooling rates, which is 5 °C/min produced crystals with the lowest aspect ratio (11.90) and widest spread of CSD (0−1200 μm). Meanwhile, the higher aspect ratio (13.58) and narrower CSD (100−900 μm) are produced at a cooling rate of 0.2 °C/min. The differences in aspect ratio and CSD are probably due to variation of nucleation event at different cooling rates. The nucleation process is typically dominant at higher cooling rate due to fast supersaturation generation, thus produce crystals with wider CSD [4, 10]. Although the aspect ratio illustrated in Figure 1 (b) is decreasing with the increase of cooling rate, a broad CSD illustrated in Figure 2 will cause a problem during the filtration process. Hence, not suitable for the crystallization process[11, 12].

Liu and co-workers reported that the cooling rate cannot be too high, to avoid uncontrolled nucleation and changes in the crystals shape [4]. Therefore, a slow cooling rate, which is either 0.2 °C/min or 0.5 °C/min seems to be more suitable for the crystallization process. As seen in Figure 1 (b), Figure 2 (a) and Figure 2 (b), the aspect ratio and CSD do not vary too much with the change of cooling rate from 0.2 °C/min to 0.5 °C/min. A cooling rate of 0.2 °C/min, however, is too slow, leading to a long batch time. From the industrial point of view, a long batch time is not preferable due to high operational cost [13]. Hence, 0.5 °C/min was selected as the best cooling rate for crystallization of mefenamic acid in ethyl acetate. Igarashi and co-workers reported that the change of cooling rate also effects the final polymorphic form and shape of the glycine crystals [14]. However, for mefenamic acid system, the crystals shape remained needle-like as illustrated in Figure 2

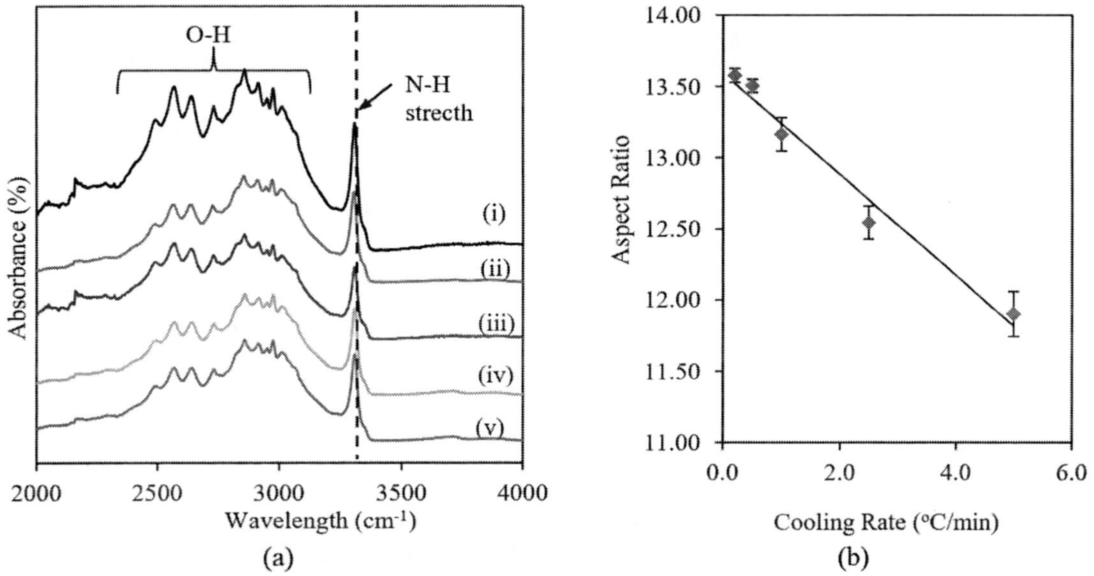

Figure 1. (a) Partial FTIR spectrum of mefenamic caid crystals obtained at cooling rate of a) 0.2 °C/min; (b) 0.5 °C/min; (c) 1.0 °C/min; (d) 2.5 °C/min; and (e) 5.0 °C/min; and (b) Effects of cooling rate on crystals aspect ratio. The solution concentration and crystallization time were fixed at 1.81 g mefenamic acid/ 100 g ethyl acetate and 30 min, respectively.

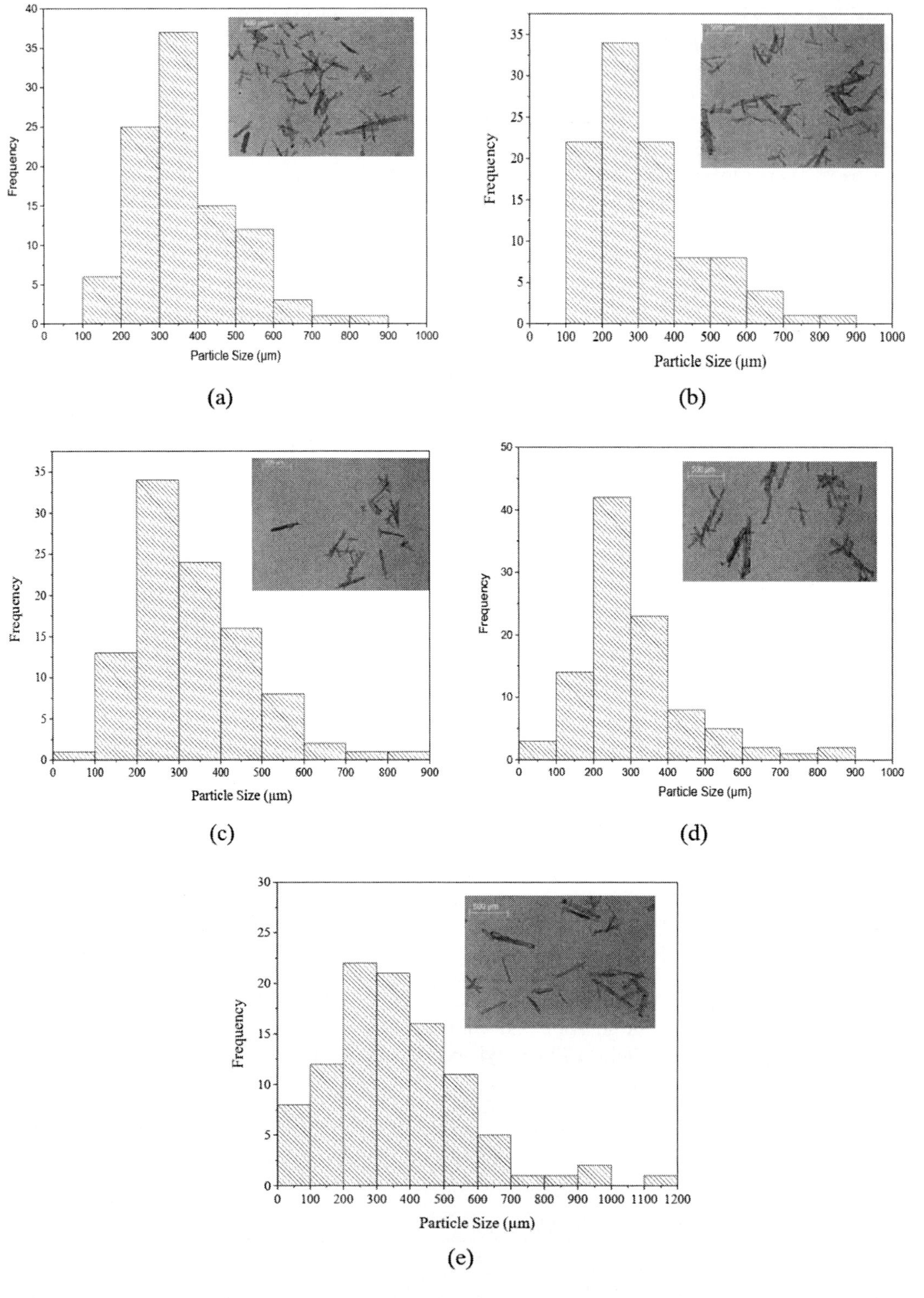

Figure 2. A CSD for different cooling rates: (a) 0.2 °C/min; (b) 0.5 °C/min; (c) 1.0 °C/min; (d) 2.5 °C/min; and (e) 5.0 °C/min. The solution concentration and crystallization time were fixed at 1.81 g mefenamic acid/ 100 g ethyl acetate and 30 min, respectively.

4. Conclusion

The crystallization of mefenamic acid were performed in batch cooling crystallization process at different cooling rates. The cooling rates show significant effect on aspect ratio and crystal size distribution. The crystals with high aspect ratio and narrower CSD (100–900 μm) was obtained at cooling rate of 0.5 °C/min. Meanwhile, the morphology and polymorphic form of mefenamic acid are remained as needle-like and Form I, respectively. As a recommendation, this work can be extended to study the effect of other crystallization conditions such as solution concentration and time on crystals properties.

Acknowledgement
Special thanks are dedicated to University Malaysia Pahang for financial support under Research Grant Scheme RDU150359.

References

[1] Reutzel-Edens SM. Achieving polymorph selectivity in the crystallization of pharmaceutical solids: Basic considerations and recent advances. Current Opinion in Drug Discovery & Development. 2006;99:806-15.
[2] Tung H-H. Industrial perspectives of pharmaceutical crystallization. Organic Process Research & Development. 2012;17(3):445-54.
[3] Heinz A, Strachan CJ, Gordon KC, Rades T. Analysis of solid-state transformations of pharmaceutical compunds using vibrational spectroscopy. Journal of Pharmaceutical Pharmacology. 2009;61:971-88.
[4] Liu JJ, Ma CY, Hu YD, Wang XZ. Effect of seed loading and cooling rate on crystal size and shape distributions in protein crystallization—a study using morphological population balance simulation. Computers & Chemical Engineering. 2010;34(12):1945-52.
[5] Kempkes M, Vetter T, Mazzotti M. Monitoring the particle size and shape in the crystallization of paracetamol from water. Chemical Engineering Research and Design. 2010;88(4):447-54.
[6] Romero S, Escalera B, Bustamante P. Solubility behavior of polymorphs I and II of mefenamic acid in solvent mixtures. Int J Pharm 1999;178:193-202.
[7] Panchagnula R, Sundaramurthy R, Pillai O, Agrawal S. Solid-state characterization of mefenamic acid. J Pharm Sci. 2004;93:1019-29.
[8] SeethaLekshmi S, Guru Row TN. Conformational polymorphism in a non-steroidal anti-inflamatory drug mefenamic acid. Crystal Growth & Design 2012;12:4283-9.
[9] Park K, Evans JMB, Myerson AS. Determination of solubility of polymorphs using differential scanning calorimetry. Cryst Growth Des. 2003;3:991-5.
[10] Pöllänen K, Häkkinen A, Reinikainen SP, Louhi-Kultanen M, Nyström L. A study on batch cooling crystallization of sulphathiazole: process monitoring using ATR-FTIR and product characterization by automated image analysis. Chemical Engineering Research and Design. 2006;84(1):47-59.
[11] Aamir E, Nagy ZK, C.D. R. Optimal seed recipe design for crystal size distribution control for batch cooling crystallisation processes. Chemical Engineering Science. 2010;65(11):3602-14.
[12] Narducci O, Jones A, Kougoulos E. Crystal product engineering in the seeded cooling crystallization of adipic acid from aqueous solution. Organic Process Research & Development. 2011;15(5):974-80.
[13] Lin R, Woo MW, Selomulya C, Lu J, Chen XD. Controlling the size of taurine crystals in the cooling crystallization process. Industrial & Engineering Chemistry Research. 2013;52(37):13449–58.
[14] Igarashi K, Sasaki Y, Azuma M, Noda H, Ooshima H. Control of polymorphs on the crystallization of glycine using a WWDJ batch crystallizer. Engineering in Life Sciences. 2003;3(3):159-63.

IOP Conf. Series: Materials Science and Engineering **205** (2017) 012026 doi:10.1088/1757-899X/205/1/012026

Effect of pH, Dosage and Concentration on the Adsorption of Congo Red onto Untreated and Treated Aluminium Dross

Mohamad Zulfika Hazielim b Zakaria[a], Rubiyah Baini[b], Nur Syuhada Ahmad Zauzi[c],

[abc]Universiti Malaysia Sarawak, Chemical Engineering Department, Faculty of Engineering, Malaysia
[a]zmzhazielim@unimas.my; [b]aznsyuhada@unimas.my; [c]ruby@unimas.my

Abstract. The adsorption of congo red onto aluminium dross was studied in batch process. The objective of this study is to adsorption capacity between untreated and treated aluminium dross in the removal of congo red. Aluminium dross was leached with 250 ml of 1% of NaOH and and precipitated with 30% H_2O_2. The treated aluminium dross being calcined at 600^oC for 3 hours. The surface area for untreated and treated aluminium dross was 10.06 m2/g and 79.80 m2/g respectively. Then the adsorption process was carried out on an orbital shaker at 200 rpm for 4 hours. In the effect of pH, it was found that untreated removes more congo red compared to the treated while in the effect of concentration solution and dosage of adsorbent, treated aluminium dross removes more congo red. In conclusion, this adsorbent was found to be effective and economically viable in the removal of congo red in waste water treatment.

1. Introduction

Aluminium dross is a solid waste from aluminium smelting industry. It is an unavoidable by products that formed on the surface of molten aluminium. According to David & Kopac, (2012), dross can be classified into three categories which is based on the metal content. The three categories are white dross, black dross and salt cake. Hwang et al., (2006) stated that white dross contain 15-70% of recoverable metallic aluminium, while black dross contain 12-18% of recoverable metallic aluminium and salt cake contains only 3-5% residual metallic aluminium. According to Sultana et al., (2013) aluminium dross contain Al_2O_3, MgF2, SiO2 and MgO.

According to Petavratzi, (2007), 95% of it usually being disposed of in a landfill. This could create the environmental problems because of landfill will act as a reactor where complex chemical reaction will occur. According to David & Kopac, (2012), when buried aluminium make contact with water with pH greater than 9 in a landfill, it will create an exothermic reaction that can retard the microbial activity in the soil. According to Murayama et al., (2012), some of this waste recycled as a deoxidizer in steel making and Petavratzi, (2007) stated that this waste can be used as a filler in a concrete. Industries looked at this as a problem due to the management cost of this waste and improper management of this waste could create the environmental problem. In this study, aluminium dross will be the raw material as an adsorbent that can remove of congo red dye in waste water.

Dye is a major pollutants that contributes in a wastewater pollution. Industries such as textiles, printing, plastics and food colouring are the common industries that pollutes wastewater with pollution. It was estimated that 106 from 107 kg of dye end up in the effluent streams worldwide Debnath et al., (2014). Liu et al.,(2015)and Namasivayam & Kanchana, (1993) stated that some of dyes hard to treat in wastewater due to the resistance for biological treatment. Mane & Babu, (2013) stated that congo red it is a benzidine based anionic disazo dye with stabile structure that make it hard to biodegrade.

IOP Conf. Series: Materials Science and Engineering **205** (2017) 012026 doi:10.1088/1757-899X/205/1/012026

According to Namasivayam & Kanchana, (1993), the colour could remain for a long distances in streams, reduce the re-aeration capacity of the stream and retards the photosynthetic activity. Debnath et al., (2014) stated that when react with water, it can metabolize to benzidine which is carcinogen to human. Membrane separation, photocatalysis and adsorption are the methods that has been developed to overcome this problem.

Zhang & Jia, (2016)and Mall et al., (2005) mentioned that adsorption is the best method since it is a simple method, low cost and stabile in the treatment. There are various methods which have been employed for removal of dyes from effluent for example activated carbon (Mane & Babu, 2013). Most of these methods are expensive. The good adsorbent is an adsorbent that have a large capacity in the removal of dyes, inexpensive and do not required pre-treatment before being used that makes the adsorption process more economically. Some of the adsorbents has been discovered from waste such as fly ash (Acemioglu, 2004), bentonite (Toor et al., 2015), saw dust (Mane & Babu, 2013), bottom ash (Saleh et al., 2012)and banana pith (Namasivayam & Kanchana, 1993)

This study is to explore the possibility of utilizing aluminium treated with sodium hydroxide for the removal of hazardous congo red from aqueous solution in a batch experiment. The effect of factors such as adsorbent dosage, contact time, initial concentration and initial pH was investigated.

2. Experimental Methods

2.1. Activation of Aluminium Dross and Surface Area Determination

Aluminium dross was collected from aluminium smelting industry, Press Metal Company located in Samalaju Industrial Park, Bintulu, Sarawak. At first, aluminium dross was leached with 250 ml of 1% concentration of sodium hydroxide and stirred for 15 minutes. After 15 minutes, samples was filtered and precipitate with 250 ml of 30% hydrogen peroxide and stirred for 30 minutes. For the last step, aluminium dross filtered and washed for several times before being calcined at 600°C for 3 hours.

The surface area of raw and treated aluminium dross were determined using BET equation to nitrogen gas adsorption (Quantachrome Instruments Co. Ltd.). The chemical composition of raw and treated aluminium dross were analysed using X-Ray Fluorescence spectroscopy equipment (Bruker S4 EXPLORER X-Ray Fluorescence). The efficiency and effectiveness of raw and treated aluminium dross as an adsorbents was measured on the adsorption of congo red.

2.2. Preparation of Congo Red

1000 mg/L (stock solution) was prepared by dissolving 1 g of congo red in 1 Liter of distilled water. 20, 40, 70 and 100 ppm solution prepared by dilution from the stock solution.

2.3. Batch Study Experiment

A batch experiment was conducted in 100 ml of congo red solution. The parameters involved in this study was adsorbent dosage, pH and initial concentration. The adsorbent dosage was varied from 0.1 to 1.0 gram, the pH of the solution was varied from pH 6 to 9 and the initial concentration of congo red was varied from 20 ppm to 100 ppm. The pH was adjusted with 0.1 M HCl and 0.1 M NaOH. Few drops of 0.01 M of NaCl was used to maintained the ionic strength of the solution. The temperature of the process maintained at room temperature. For the samples measurement, 10 minutes interval within the first hour, 30 minutes for the second hour and subsequent 1 hour sampling until the adsorption process reach equilibrium. The final concentration of congo red solution was measured using UV-Vis spectrophotometer with 500 nm wavelength absorbance. The amount of congo red adsorbed by aluminium dross was calculated according to equation 1:

$$q = \frac{C_0 - C_e}{m} V$$

q (mg/g)= amount of congo red adsorbed per unit gram of aluminium dross; C_0 and C_e (mg/l) is the initial and final concentration of congo red; V (l) is the solution volume and m (g) is the dosage of aluminium dross.

IOP Conf. Series: Materials Science and Engineering **205** (2017) 012026 doi:10.1088/1757-899X/205/1/012026

3. Results and Discussion

3.1. *Surface area of aluminium dross*

The untreated and treated aluminium dross being analysed with nitrogen adsorption-desorption measurements to determine the surface area of both samples. It was found that the surface area for untreated and treated aluminium dross was 10.06 m^2/g and 79.80 m^2/g respectively.

3.2. *Effect of pH on the adsorption of congo red onto untreated and treated aluminium dross*

Table 1. Effect of pH on the adsorption of congo red onto 0.1 gram of untreated and treated aluminium dross

	Percentage removal (%)	
	Untreated aluminium dross	Treated aluminium dross
pH 6	51.45	82.51
pH 8	65.62	82.05
pH 9	97.58	67.21

From table 1, it shows that for the untreated aluminium dross, the percentage removal increases as the pH increases while for the treated aluminium dross, the percentage removal decreases as the pH increases. According to Bhat et al., (2015), the solubility of metal ions, concentration of counterions on the functional groups of adsorbent and the degree of the ionization of the adsorbate were affected by pH. Aluminium dross majority comprises of Al_2O_3 that creates a positive charged when dissolved in water. Therefore, electrostatic attraction occur between the positively charged of treated aluminium dross with negatively charged of congo red dyes. When the pH of congo red solution increases, it shows that the adsorption of congo red decreases due to the positive charge on the active sites of aluminium dross decreases. It is because when the pH increases, more –OH ions will released and there is a competition between the –OH ions and the anionic of congo red dye for the binding process on the positive charge active sites on aluminium dross. According to Kosmulski, (2001), the pH point zero charge (pHpzc) for alumina was pH 8.6. When the pH of congo red solution above 8.6, the surface of aluminium dross will become a negative charge and repulsion of electrostatic charge between congo red and aluminium dross will occur. While, for the untreated aluminium dross, the percentage removal increases as the pH increases. This is because of the positive charged on the active sites of untreated aluminium dross was no effected by pH. In addition, it contains sufficient of positive charged sites for the removal of 20 ppm of congo red concentration.

3.3. *Effect of concentration on the adsorption of congo red onto untreated and treated aluminium dross*

Table 2. Effect of concentration of congo red on the adsorption onto 0.1 gram of untreated and treated aluminium dross

	Percentage removal (%)	
	Untreated aluminium dross	Treated aluminium dross
20 ppm	51.45	82.51
40 ppm	29.36	69.22
70 ppm	25.20	67.19
100 ppm	19.52	42.34

Table 2 shows the removal of different concentration on congo red onto 0.1 gram of untreated and treated aluminium dross at pH 6 solution. Both of the adsorbent shows that the percentage removal decreases as the concentration increases. This is because of the availability of the active sites of the adsorbent. When the concentration increases, the negative ions of dye solution dissociated also increases. The availability of active sites for the removal of congo red are limited for the high concentration of congo red.

3.4. Effect of dosage of adsorbent on the adsorption of congo red onto untreated and treated aluminium dross

Table 3. Effect of dosage of adsorbent on the adsorption of 20 ppm concentration of congo red at pH 6

	Percentage removal (%)	
	Untreated aluminium dross	Treated aluminium dross
0.1 g	51.45	82.51
0.4 g	80.68	88.90
0.6 g	92.56	69.90
0.8 g	94.36	90.73
1.0 g	96.44	90.96

The table shows that percentage removal increases as the adsorbent dosage increases. Increase in the dosage of adsorbent, it will promote more active sites that participates in the adsorption of congo red. Bhat et al., (2015) stated that when the dosage increases, the competition between the anionic dyes for binding sites decreases. From the result, it was found that the highest removal of congo red was the 1.0 gram of adsorbent dosage with high surface aluminium dross.

4. Conclusion

In this study, the adsorption of congo red was conducted with untreated and treated aluminium dross. The surface area for untreated and treated aluminium dross was 10.06 m^2/g and 79.8 m^2/g respectively. In the effect of pH, it was found that untreated remove high percentage of congo red compared to the treated aluminium dross. While in the effect of initial concentration and adsorbent dosage, treated aluminium dross remove high percentage of congo red compared to the untreated aluminium dross. From this study, it can be concluded that aluminium dross that has no value can be effectively used for the adsorption of congo red.

5. Acknowledgement

The authors acknowledge the research grant provided by the UNIMAS under Ministry of Higher Education Fund grant (RAGS/c(14)/948/2012(49).

References

[1] Acemioglu, B. (2004). Adsorption of Congo red from aqueous solution onto calcium-rich fly ash, 274, 371–379. http://doi.org/10.1016/j.jcis.2004.03.019

[2] Bhat, A., Megeri, G. B., Thomas, C., Bhargava, H., Jeevitha, C., Chandrashekar, S., & Madhu, G. M. (2015). Adsorption and optimization studies of lead from aqueous solution using g -Alumina. Biochemical Pharmacology, 3(1), 30–39. http://doi.org/10.1016/j.jece.2014.11.014

[3] David, E., & Kopac, J. (2012). Hydrolysis of aluminum dross material to achieve zero hazardous waste. Journal of Hazardous Materials, 209–210, 501–509. http://doi.org/10.1016/j.jhazmat.2012.01.064

[4] Debnath, S., Maity, A., & Pillay, K. (2014). Impact of process parameters on removal of Congo red

by graphene oxide from aqueous solution. Biochemical Pharmacology, 2(1), 260–272. http://doi.org/10.1016/j.jece.2013.12.018

[5] Hwang, J. Y., Huang, X., & Xu, Z. (2006). Recovery of Metals from Aluminum Dross and Saltcake. Current, 5(1), 47–62.

[6] Kosmulski, M. (2001). Evaluation of point zero charge of aluminium oxide reported in the literature.

[7] Liu, X., Niu, C., Zhen, X., Wang, J., & Su, X. (2015). Journal of Colloid and Interface Science Novel approach for synthesis of boehmite nanostructures and their conversion to aluminum oxide nanostructures for remove Congo red. Journal of Colloid And Interface Science, 452, 116–125. http://doi.org/10.1016/j.jcis.2015.04.037

[8] Mall, I. D., Srivastava, V. C., Agarwal, N. K., & Mishra, I. M. (2005). Removal of congo red from aqueous solution by bagasse fly ash and activated carbon : Kinetic study and equilibrium isotherm analyses, 61, 492–501. http://doi.org/10.1016/j.chemosphere.2005.03.065

[9] Mane, V. S., & Babu, P. V. V. (2013). Journal of the Taiwan Institute of Chemical Engineers Kinetic and equilibrium studies on the removal of Congo red from aqueous solution using Eucalyptus wood (Eucalyptus globulus) saw dust. Journal of the Taiwan Institute of Chemical Engineers, 44(1), 81–88. http://doi.org/10.1016/j.jtice.2012.09.013

[10] Murayama, N., Maekawa, I., Ushiro, H., Miyoshi, T., Shibata, J., & Valix, M. (2012). International Journal of Mineral Processing Synthesis of various layered double hydroxides using aluminum dross generated in aluminum recycling process. International Journal of Mineral Processing, 110–111, 46–52. http://doi.org/10.1016/j.minpro.2012.03.011

[11] Namasivayam, C., & Kanchana, N. (1993). Removal of Congo Red from Aqueous Solution by Waste Banana Pith, 1(1), 33–42.

[12] Petavratzi, E. (2007). Residues from aluminium dross recycling in cement, (November), 1–8.

[13] Saleh, S. M., Maarof, H. ., Rahim, S. N. S. ., & Nasuha, N. (2012). Adsorption of Congo red onto Bottom Ash.

[14] Sultana, U. K., Gulshan, F., Gafur, M. a, & Kurny, a S. W. (2013). Kinetics of Recovery of Alumina from Aluminium Casting Waste through Fusion with Sodium Hydroxide, 1(3), 30–34. http://doi.org/10.12691/materials-1-3-1.

[15] Toor, M., Jin, B., Dai, S., & Vimonses, V. (2015). Journal of Industrial and Engineering Chemistry Activating natural bentonite as a cost-effective adsorbent for removal of Congo-red in wastewater. Journal of Industrial and Engineering Chemistry, 21, 653–661. http://doi.org/10.1016/j.jiec.2014.03.033.

[16] Zhang, Y., & Jia, Y. (2016). Preparation of porous alumina hollow spheres as an adsorbent for fl uoride removal from water with low aluminum residual. Ceramics International, 42(15), 17472–17481. http://doi.org/10.1016/j.ceramint.2016.08.052.

IOP Conf. Series: Materials Science and Engineering **205** (2017) 012027 doi:10.1088/1757-899X/205/1/012027

Optical and Morphological Characterization of Sonochemically Assisted Europium Doped Copper (I) Oxide Nanostructures

J A M Cosico, P K Ruales and M C Marquez

School of Chemical Engineering and Chemistry, Mapua Institute of Technology Manila Philippines
menandro.marquez@gmail.com

Abstract. In the age where application of nanotechnology in our society has proven to be eminent, different routes of synthesizing nanoparticles have emerged. In this study nanoparticles of cuprous oxide (Cu_2O) doped with different amounts of europium was prepared by using solution precursor route approach with the aid of ultrasonic sound. Copper sulphate and europium (III) nitrate pentahydrate was used as source for copper ions and europium ions respectively. X-ray diffraction (XRD) and Fourier Transform Infrared spectroscopy (FTIR) were used to elucidate the cubic crystal structure and organic impurities present on Cu_2Onanoparticles. UV-Vis spectroscopy was used to determine the absorption spectrum of the nanoparticles in the wavelength range of 400nm to 700nm. The bandgap of the undoped and doped Cu_2O were found to fall between 2.1eV - 2.3eV. Scanning Electron Microscopy (SEM) coupled with energy dispersive x-ray was used to observe the dendritic and rodlike morphology and the presence of europium in the synthesized Cu_2O nanoparticles. The observed effect on the absorbance of Cu_2O upon adding Eu and a facile way of synthesizing Cu_2O nanoparticles could bring a positive impact on the production of functional devices for optoelectronic and energy applications.

1. Introduction

Metal oxide nanostructures have profound properties in different fields like optics, optoelectronics, catalysts, biosensors etc. Copper oxides are known for their good room temperature mobility and high minority carrier diffusion length [1,3, and 11]. It is known to have a direct bandgap energy ranging from ~ 1.8eV – 2.3eV with its high absorption in the visible region. [5] Knowing this, copper oxide-based materials have been widely investigated due to their potential applications in chemical, photochemical and electrochemical fields [4, 11].

Cuprous oxide or Copper (I) oxide (Cu_2O) thin films can be prepared by several methods such as electrodeposition, sonochemical method, thermal relaxation liquid phase reduction, thermal evaporation, sol-gel, etc. Among these methods, sonochemical assisted techniques will be utilized for preparation of the Cu_2O in powdered form. Sonochemical processing has been recognized as a useful technique for making nanoparticles, nanoemulsions, nanocrystals, biofuels, etc. Sonication is commonly used in nanotechnology for evenly dispersing nanoparticles in liquids. Ultrasound causes chemical reactions through the process of acoustic cavitation – the formation, growth, and implosive collapse of gas bubbles in a liquid. [1].

In comparison with other thin films of oxide semiconductors such as In_2O_3, SnO_2, and ZnO [2,6], few data are available on the synthesis of europium (Eu) doped cuprous oxide (Cu_2O) films. With Cu_2O known for its direct bandgap energy and its potential application as solar light absorbers in photovoltaic devices, this study explores changes on Cu_2O in terms of optical and morphological properties upon adding Eu.

2. Methodology

2.1 Materials

The chemicals used in this study were copper sulphate pentaydrate (Sigma-Aldrich). Sodium dodecyl sulphate, ascorbic acid, sodium hydroxide and europium nitrate pentahydrate were used as received and without further purification.

2.2 Synthesis of Doped and Undoped Cu_2O

Nanoparticles of Eu-doped Cu_2O were prepared by using solution precursor route approach with the aid of ultrasonic sound. An aqueous solution of 0.01 M Copper sulphate pentahydrate ($CuSO_4 \cdot 5H_2O$) then, 0.1 M of sodium dodecyl sulphate) solution was added to copper solutioe which formed solution A via sonochemical dissolution for 20 minutes. Aqueous Europium nitrate pentahydrate ($Eu(NO_3)_3 \cdot 5H_2O$) was prepared with varying concentration. (0.5%, 1%, 2%, 3% and 5%) and was added to solution A. It was mixed with the aid of ultrasonic sound for 20 minutes and was called solution B. 0.5 M Ascorbic acid ($C_6H_8O_6$) and 1.0 M Sodium hydroxide (NaOH) was added to solution B and the final solution was aged for 24 hours at room temperature. Afterwards, it was filtered and then vacuum dried for three hours at room temperature to produce the doped and undoped Cu_2O.

The study is guided with the aid of XRD (X-ray Diffraction) and FTIR (Fourier Transform Infrared Spectroscopy). After confirming the crystallographic structure of the undoped and Eu-doped Cu_2O, SEM (Scanning Electron Microscope) and EDX (Energy Dispersive X-ray Spectroscopy) are used to confirm the formation of nanostructures and confirm that Eu^{3+} is incorporated to the Cu_2O formed. The samples are then subjected to UV/Vis (Ultraviolet-Visible Spectroscopy) to confirm that upon addition of Eu^{3+}, optical absorption of Cu_2O nanostructures is altered or enhanced.

3. Results and Discussion

The study focused on the effects of Europium (Eu^{3+}) when doped to Copper (II) Oxide (Cu_2O). Moreover, the study highlights the effects of Eu^{3+} as a dopant on the optical and morphological properties of Cu_2O.

Pristine Cu_2O and Eu-doped Cu_2O nanoparticles were synthesized using solution precursor approach aided with ultrasonic sound. When copper sulphate pentahydrate is dissolved in water, ions of Cu are formed (Cu^{2+}) which will then be reduced by ascorbic acid. With the addition of NaOH as a source of hydroxide ion (OH^-), copper hydroxide is formed ($Cu(OH)_2$) which can eventually be transformed to Cu_2O. During this process, some Eu^{3+} will replace Cu^{2+} simultaneously, in the formation of Cu_2O, as a substitutional impurity to form Cu_2O doped with Eu^{3+}. It is known that sonochemical method is used in the experiments to incorporate Eu^{3+} to metal oxides.

The X-ray diffraction pattern of the synthesized sample is shown in Figure 1. The sample used for the XRD was the one with 5% of Eu^{3+} loaded on to Cu_2O. Looking at the diffractogram presented in Figure 1 it was observed that the synthesized Eu-Cu_2O is crystalline in nature with peaks located at 2Θ equal to 36.54°, 42.44° and 61.58° each corresponding to planes indexed to (111), (200) and (220) respectively.

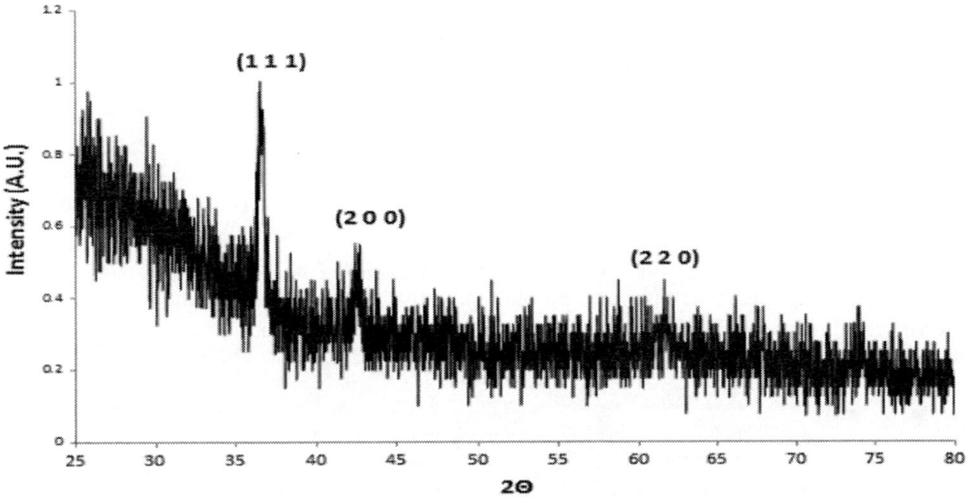

Figure 1. XRD diffractogram of doped Cu2O

These peaks and indexed planes are in accordance to JCPDS 05–0667. These peaks are also the characteristic peaks of Cu_2O with cubic structure. No other phase, such as europium oxide and other forms of copper oxide were found which indicate the formation of crystalline Cu_2O phase only.

Table 1.2a and 1.2b show the elements present on the undoped and doped samples. EDX was used to determine the incorporation of Eu^{3+} on Cu_2O. A value of -2.58 in the undoped Cu_2O signifies that undoped Cu_2O does not contain Eu^{3+}. This negative sign also implies that the signal for Eu detected in the sample is even lower than that of the background. However, a value +0.94 in Table 1.2b mean that indeed Eu^{3+} was incorporated in the Cu_2O crystal. Figure 5 shows the EDX spectra summarized in Tables 1 a and 1b. The resolution limit of the EDX system that was used in the study is 1000 ppm, equivalent to 0.1 wt%.

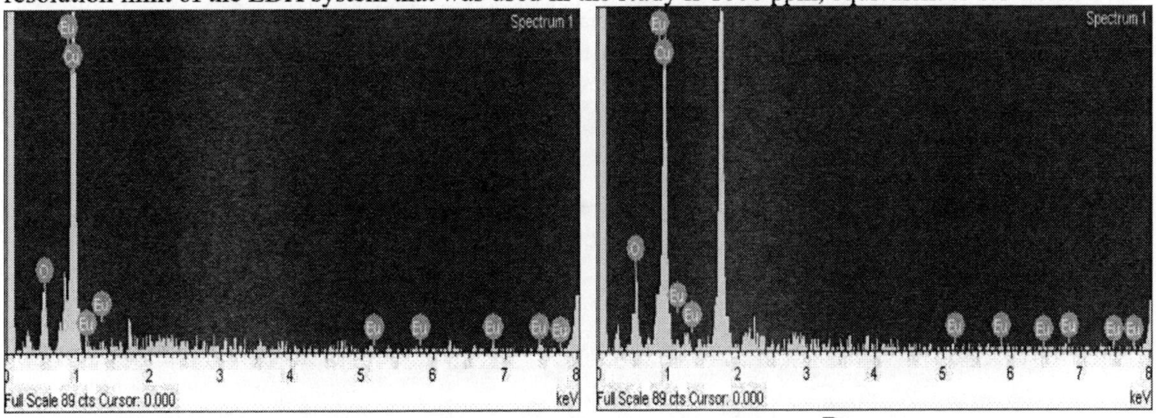

A B

Figure 2. EDX of (A)undoped Cu_2O and (B)Eu doped Cu_2O

Table 1a. EDX Results – Elemental analysis of undoped Cu_2O.

ELEMENT	Weight %	Atomic %
O	13.26	37.38
Cu	89.32	63.38
Eu	-2.58	-0.77

Table 1b. EDX Results – Elemental analysis of doped Cu_2O.

ELEMENT	Weight %	Atomic %
O	18.32	47.27
Cu	80.74	52.47
Eu	0.94	0.26

In this paper, SEM was used to investigate the effect of adding Eu^{3+} ions on the morphology of Cu_2O. This is then paired with EDX to confirm the incorporation of Eu^{3+}. These images are illustrated in Figure 3. It can be observed that in this micrograph, aggregates of copper oxide nanoparticles are evident. Voids, canyons and formation of dendritic structures are also evident, signifying a porous microfilm can be produced when it is casted on a substrate. Figure 3 shows the low magnification and high magnification (insets) SEM image of the samples with and without europium.

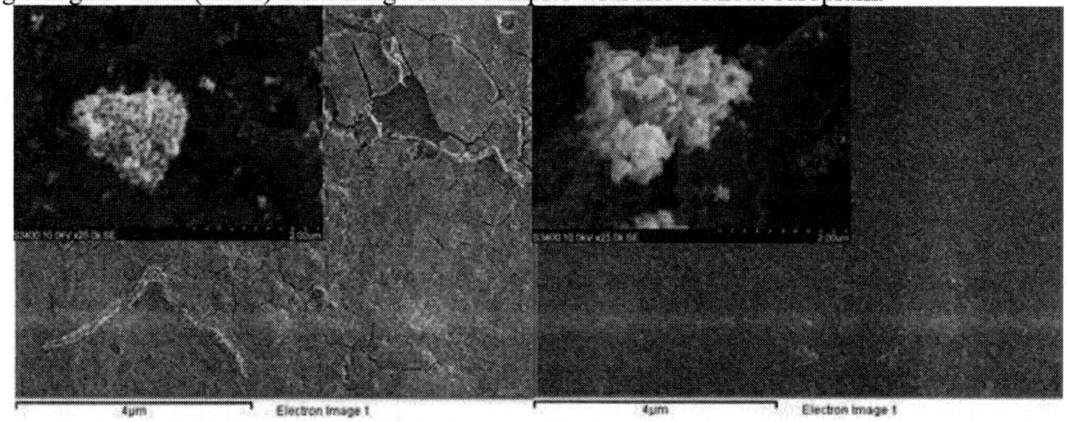

Figure 3. Low magnification of SEM results for left) undoped Cu_2O and right0 Eu-doped Cu_2O Insets (25000x magnification scale bar = 2um)

The SEM images obtained at 15,000X magnification are shown in Figures 4a – 4f . At 0.5%wt Eu doped to Cu_2O, as represented Figure 5b the particles are more dispersed, in a dendritic manner, and particles become partly smaller in size with respect to the images obtained from the undoped Cu_2O.

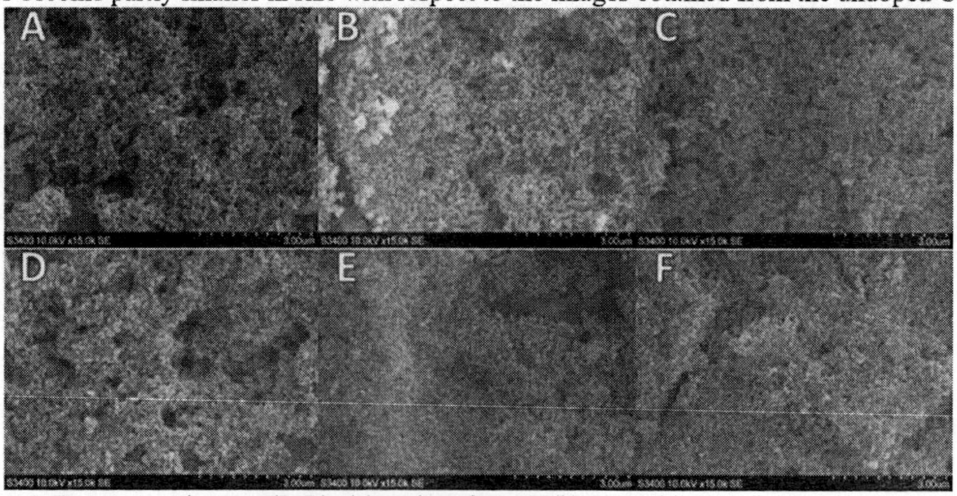

Figure 4. From top to bottom (L-R): (a) undoped Cu_2O (b) 0.5% (c) 1% (d) 2% (e) 3% (f) 5% Eu-doped Cu_2O

At 1%wt Eu doped to Cu_2O, as represented by Figure 4c, voids are visible, but particles are more uniformly sized and are dispersed, still in a dendritic manner, more evenly compared to the undoped Cu_2O. Although a better distribution of particles, agglomeration of particles can be observed, edges of agglomerated particles become distinct forming rod like structures as confirmed by the image shown in the inset image in Figure 3.
UV/Vis was used in this study to compare the effect of the different concentrations of Eu^{3+} on the optical bandgap of doped Cu_2O with respect to that of the undoped Cu_2O. The obtained absorbance spectra of the undoped and Eu-doped Cu_2O are shown in Figure 5. The summary of the computed optical band gap of the doped and doped samples is listed in Table 2. It should be noted that the listed wavelengths in Table correspond to the onset of absorption for each sample. Based on the data gathered from the UV/Vis, in can be infer that as the amount of Eu^{3+} concentration increases, the E_g decreases, which means that addition of Eu^{3+} favors higher absorbance of visible light. Concentrations of Eu^{3+} ranging from 0.5% to 3% gave a direct linear relationship to the optical bandgap of Cu_2O lower than that of the undoped Cu_2O. This decreased in bandgap to 2.13eV makes it favorable for its potential application as visible light absorbers in photovoltaic devices..

Table 2. Onset of absorption and approximated band gap of doped and undoped Cu_2O.

SAMPLE	Wavelength (nm)	E_g (eV)
Undoped	552	2.25
0.5% Eu-	583	2.13
1.0% Eu-	577	2.15
2.0% Eu-	564	2.20
3.0% Eu-	558	2.22
5.0% Eu-	548	2.26

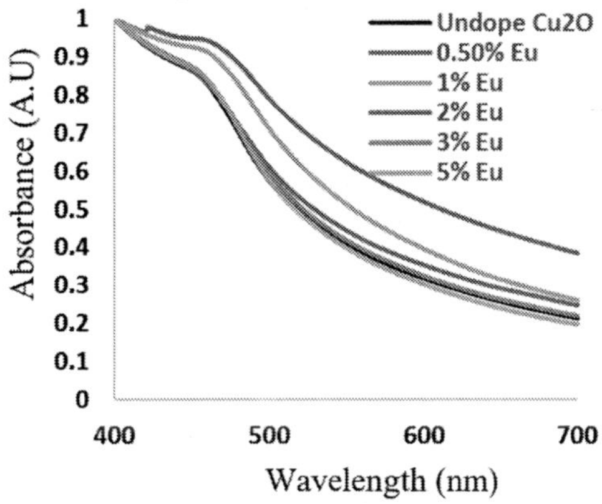

Figure 5. UV-Vis of doped and undoped Cu_2O

4. Conclusion

Undoped and Eu-doped Cu_2O nanoparticles were successfully synthesized via solution precursor route with the aid of ultrasonic sound. The incorporation of Eu^{3+} on Cu_2O did not alter the crystal structure and no formations of other species of copper and europium were observed. The SEM images revealed dendritic nanostructures composed of agglomerated nanoparticles. Optical characterization together with the calculation of bandgap confirmed that the synthesized Cu_2O nanoparticles are absorbing in the visible region. Moreover, the incorporation of Eu^{3+} on Cu_2O decreased the bandgap from ~ 2.25 eV to ~ 2.13 eV. Subsequently, further addition of Eu^{3+} resulted in the increased in bandgap of ~2.28eV. Further studies can be done on the effect of loading more Eu^{3+} on Cu_2O and determine if it could increase its bandgap at higher Eu loading.

Acknowledgment
The authors would like to thank the Philippine Council for Industry, Energy and Emerging Technology Research and Development - Department of Science and Technology (PCIEERD-DOST) of the Philippine Government for the financial support they have extended for this study.

References

[1] Abulizi A., Yang G H and Zhu J J 2014. One-step simple sonochemical fabrication and photocatalytic properties of Cu_2O–rGO composites. *Ultrasonics Sonochemistry*, 129-135
[2] Al Rifai S A, and Kulnitskiy B. 2013. Microstructural and optical properties of europium-doped zinc oxide nanowires. *Journal of Physics and Chemistry of Solids*, 1733-1738.

[3] Luo X, Li C , Yang D, Liu F, Chen, Y. 2015. Sonochemical synthesis of porous Cu_2O@Cu hollow spheres and their photo-catalysis. *Materials Chemistry and Physics*, 252-258
[4] Mao B G, Chu D Q, Wang L M., Wang A X, Wen Y J, and Yang X Z 2013. Ultrasound-assisted synthesis of sea urchin-like Cu_2O architectures. *Materials Letters*, 62-65.
[5] Messaoudi O, Assaker B, Gannaouni M, Souissi A, Makhlouf H B and Chtourou R. 2016. Structural, morphological and electrical characteristics of electrodeposited Cu_2O: Effect of deposition time. *Applied Surface Science*, 383-388.
[6] Panwar A, Yadav, K. 2015. A novel one-pot synthesis of hierarchical europium doped ZnO nanoflowers. *Materials Letters*, 30-34.
[7] Shui, A., Zhu, W., Xu, L., Qin, D., & Wang, Y. (2013). Green sonochemical synthesis of cupric and cuprous oxides nanoparticles and their optical properties. *Ceramics International*, 8715-8722.

IOP Conf. Series: Materials Science and Engineering **205** (2017) 012028 doi:10.1088/1757-899X/205/1/012028

Evolution of Cobalt Oxide Nanostructures on Glass Substrate via Two Step Solution Route Synthesis

A. R. Mariscal and M. C. Marquez

School of Chemical Engineering and Chemistry, Mapua Institute of Technology Manila Philippines, 0002
menandro.marquez@gmail.com

Abstract: Controlling the morphology in the nanoscale has proven to be an effective way to drastically change the properties of materials. In this study, several morphologies of cobalt oxide (Co3O4) were synthesized by employing a two-step solution route. A Co3O4 seed layer was first deposited on the glass substrate via spin coating using a cobalt acetate precursor followed by chemical bath deposition of another Co3O4 layer using cobalt nitrate precursor. The effect of the seed layer and the deposition times on the morphology of the secondary Co3O4 nanostructures was verified by employing scanning electron microscopy (SEM). Morphologies like nanoplatelets, nanorods, nanofibrils and porous nanowalls were observed in the SEM. Other characterization techniques such as Fourier transform infrared spectroscopy (FTIR) and X-ray diffraction technique (XRD) were used to elucidate the formation of Co3O4. Furthermore, the surface of the seed layer was determined via atomic force microscope. It was found that the roughness of the seed layer ranges from ~0.3 to 3 nm depending on the concentration of the cobalt precursor used. Results showed that the morphology of Co3O4 can be easily modified using the two-step solution route technique. The alterations of the morphology of Co3O4 could lead to morphologies with unique properties and development of functional materials applicable in the field of energy and electronics.

1. Introduction

Cobalt oxide is a heavily investigated metal oxide due to its versatility in terms of applications. Cobalt oxide has attracted the interest of scientist in the recent decade for it has shown superb electrochemical properties which makes a viable material in the fields of batteries, electrochemical capacitors, electrochromic devices, gas sensors, as well as catalysts for oxygen reduction reaction [1-4].

Cobalt Oxide can be deposited in an electrode via successive ionic layer adsorption and reaction (SILAR), chemical spray pyrolysis, chemical vapor decomposition, solvothermal method, electrodeposition, chemical bath deposition, chemical combustion method, redox route, and pulsed laser deposition methods[5,6]. Cobalt oxide spinel nanostructures fabricated using the synthetic methods, yielded nanostructures with various morphologies such as nanotubes, nanowires, nanoplates, and nanoparticles[7]. Among these techniques, chemical bath deposition stand out due to certain advantages. Chemical bath deposition, or chemical solution deposition, is a preferred method since it has a low cost, simple design, great potential to scale-up, and produces high quality nanorods [8].

In this study, we reported the effects of seed layer, seed layer solution concentration, and growth time on the nanostructures formed. Nano sized crystalline metal oxides are preferred due to their large

IOP Conf. Series: Materials Science and Engineering **205** (2017) 012028 doi:10.1088/1757-899X/205/1/012028

surface areas, unusual adsorptive properties, surface defects and fast diffusivities [9,10]. Co_3O_4 is a material which may yield as a clean source of energy due to its capability as a catalyst for water splitting applications.

2. Methodology

Solutions were prepared using analytical reagent grade chemicals without further purification. Cobalt (II) acetate tetrahydrate from Sigma Aldrich was used as cobalt source for the seed layer. Cobalt nitrate hexahydrate and hexamethylenetetramine (HMTA) were used for the synthesis of the cobalt oxide nanostructures.

The substrates were prepared by cutting 5 mm x 15 mm glass slides using a diamond glass cutter. The glass substrates were then cleaned in ethanol, acetone, and distilled water, using a low frequency sonicator for 5 minutes each. This standard cleaning method was used to remove the grease and other contaminants on the glass substrates which may have been deposited during the cutting phase of the glass. Solutions of 0.03M, 0.04M, and 0.05M cobalt (II) acetate tetra-hydrate dissolved in methanol were prepared for the seed layer. The seed layer was deposited with aid of a spin coater. The glass substrate was attached to the coater and spun at 2400 rpm. The substrates were then annealed at 450° C for 1 hour.

For the formation of the secondary nanostructures, 0.02M of cobalt nitrate hexahydrate and hexamine (HMTA) solution were used as precursors. Cobalt nitrate hexahydrate and hexamine (HMTA) were dissolved in deionized water with molar concentrations; and were sonicated for 30 minutes. The nanostructures were then grown onto the substrate via chemical bath deposition with temperature set at 90°C. After growing, the substrate was washed with deionized water and then annealed at 450 °C for 1 hour.

3. Results and Discussion

3.1. X-Ray Diffraction

The diffractogram shows characteristic peaks which confirmed what the crystal structure of the synthesized material. The recorded peaks of the fabricated cobalt oxide seed layer and the cobalt oxide nanostructures grown on top of the seed layer confirmed this claim and corroborates with the standard JCPDS data (JCPDS card No: 073-1701) for cubic cobalt oxide (Co_3O_4).

The formation of Co_3O_4 seed layer (equations 1-3) and nanostructures (equations 4-6) using solution route technique can be summarized in the proposed chemical reactions presented below:

$$Co(CH3COO)_2 \cdot 4H_2O + 2NaOH \rightarrow Co(OH)_2 + 2CH_3COONa + 4H2O \quad (1)$$

$$4Co(OH)_2 + O_2 + H2O \rightarrow 4Co(OH)_3 \quad (2)$$

$$Co(OH)_3\downarrow \rightarrow \text{(annealed in air at 450 °C)} \rightarrow Co_3O4 + H_2O \quad (3)$$

$$Co(NO_3)_2 \cdot 6H2O + NaOH \rightarrow Co(OH)2 + NaNO3 \quad (4)$$

$$4Co(OH)_2 + O_2 + H_2O \rightarrow 4Co(OH)_3 \quad (5)$$

$$Co(OH)_3\downarrow \rightarrow \text{(annealed in air at 450 °C)} \rightarrow Co_3O4 + H_2O \quad (6)$$

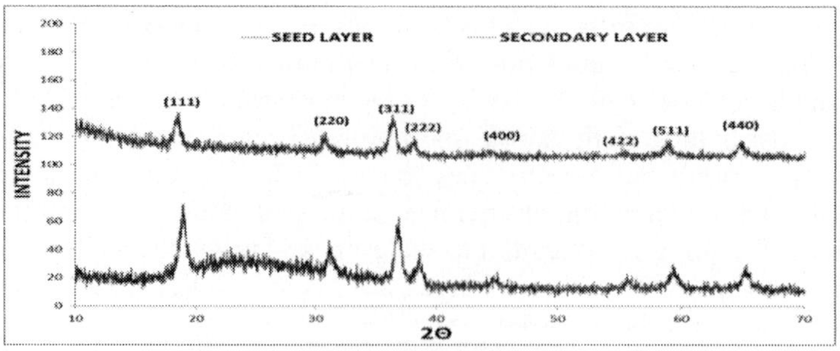

Figure.1. XRD diffractogram of cobalt oxides (Co_3O_4) (from Cobalt (II) acetate tetatrahydrate and Cobalt (II) nitrate hexahydrate precursors)

IOP Conf. Series: Materials Science and Engineering **205** (2017) 012028 doi:10.1088/1757-899X/205/1/012028

3.2 Atomic Force Microscopy

Atomic Force microscopy (AFM) was employed to observe the surface roughness of the deposited seed layer and nanostructures. Based on the results; increasing the concentration of the solution leads to increase in the surface roughness of the seed layer. This surface roughness and concentration dependence is attributed to the increase in the Van der Waals forces between the particles which is due to less distance between particles if the solution is more concentrated[11, 12].

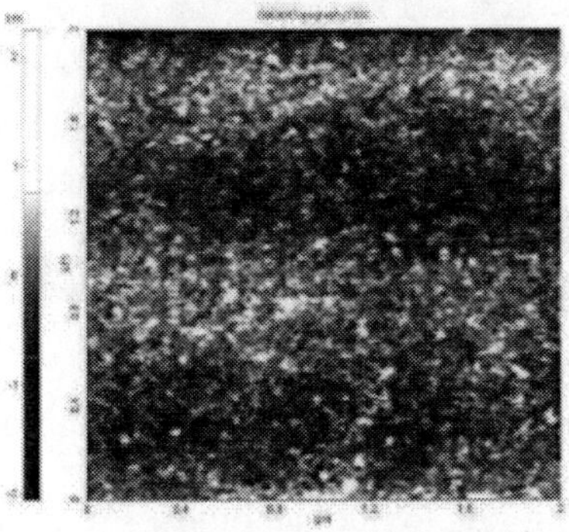

Figure.2. Atomic Force micrograph of cobalt oxide seed layer with concentration of 0.03M 2D with reported roughness of root mean square of 0.383 nm.

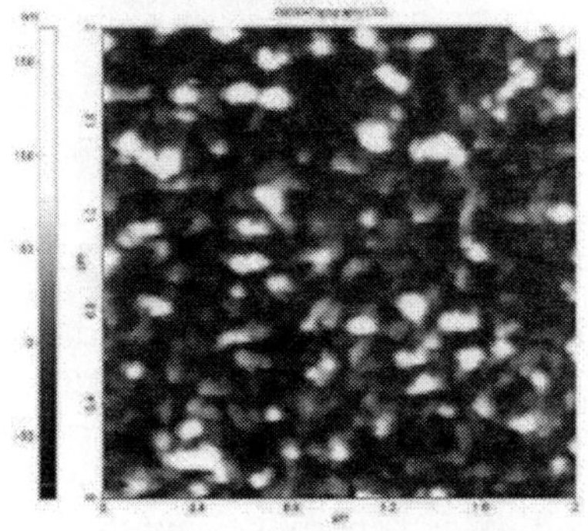

Figure.3. Atomic Force micrograph of cobalt oxide nanostructures with concentration of 0.03M with reported roughness of root mean square of 38.588 nm

3.3 Scanning Electron Microscopy

Scanning electron microscopy (SEM) was used to investigate the morphology of the seed layer as well as the influence of growth time and seed layer concentration on the nanostructures formed.

Figure.4. Seed layers with concentrations A) 0.03M, B) 0.04M, and C) 0.05M.

With the SEM images of the seed layer there is an increase in number of particles as the seed layer concentration increase. This results in a rougher film, as confirmed by the AFM micrographs. Increasing the concentration of the solution, the presence of more cobalt oxide particles means that the distance between such particles is decreased. This causes the particles to agglomerate. Hence, more particles are present on the 0.05M seed layer compared to 0.03M seed layer. The diameter of the particles present in the seed layer film was not further investigated. The dark spots in the SEM images are not cobalt oxide particles; rather they are void undeposited spaces. It was noted that there are more dark spots present in the 0.03M SEM image as compared to the 0.05M seed layer.

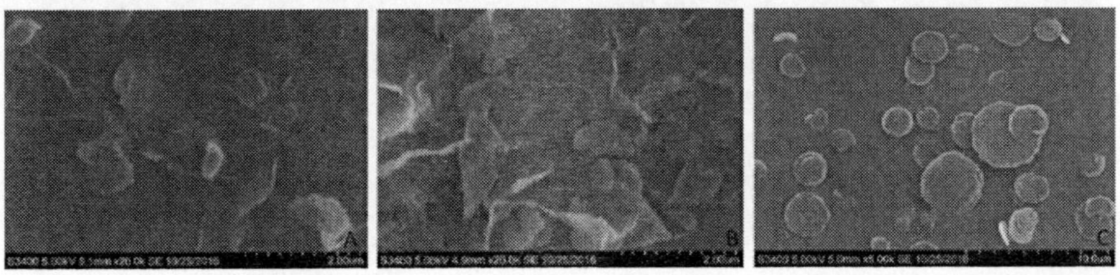

Figure.5. Nanostructures formed without seed layer, films that were grown at A) 0.5hour, B) 1 hour, C) 3 hours.

Shown in Figure 3.5 are the SEM images of one step chemical bath deposition. As observed, there are not much nanostructures formed in this method. The only nanostructure that was present in the three samples, without seed layer, was nanosheets developing into nanoplates (Fig. 3.5., Image B). The nanosheets were parallel to the substrate and, as observed. There were no dot-like particles observed, unlike the ones present on the cobalt oxide seed layer.

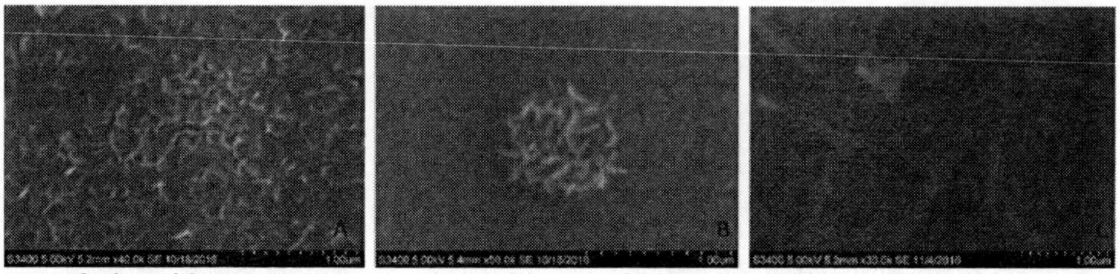

Figure.6. Cobalt oxide nanostructures on seed layer with concentration 0.03M with varying growth times, A) 0.5hour, B) hour, C) 3 hours.

The above images show the evolution of cobalt oxide fibrils to cobalt oxide nanorods. These structures had the same seed layer concentration which was 0.03M. It was confirmed the formation of cobalt oxide nanorods using the same method in fabricating zinc oxide nanorods. From the images, the trend that was observed was that in the 0.03M fibrils and rods were common structures. Observing image B, the underlying layer has a different contrast compared to the underlying layer of the first image (image A). This shows that the nanostructure formed in image B is a tertiary formation. This means that given enough time, a nanostructure will form on top of a previously grown nanostructure.

Figure.7. Cobalt oxide nanostructures on seed layer with concentration 0.04M with varying growth time, A) 0.5hour, B)1 hour, C) 3 hours.

Nanowalls were the dominating structures on the 0.04M seed layer concentration. As seen on the image C, there is a difference in contrast between the underlying layer on the image compared to the top left and top right images. This also exhibits the same tertiary growth behaviour as stated observed in the previous images.

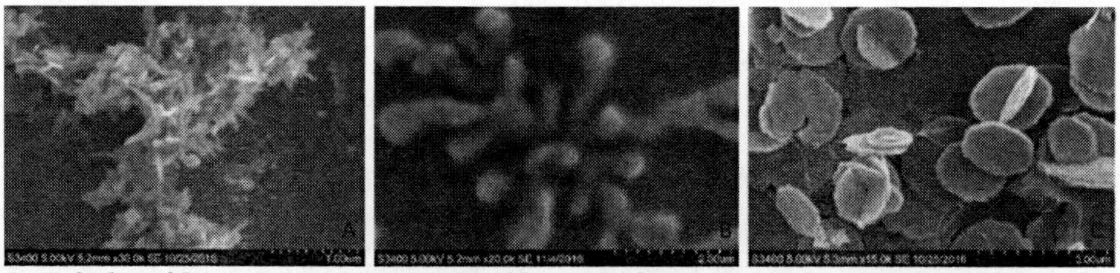

Figure.8. Cobalt oxide nanostructures on seed layer with concentration 0.05M with varying growth times, A) 0.5hour, B) 1 hour, C) 3 hours.

A coral-like nanostructure was found on 0.05M seed layer concentration grew on 1 hour. It seems to grow perpendicular from the glass substrate. Other observed nanostructures that were grown on the 0.05M seed layer were nanosheets. Unlike the nanosheets found on the no seed layer samples, these nanosheets are perpendicular to each other. There were no trends observed. However, from these set of images it was found that there were nanosheets formed. It was proven that by depositing a seed layer onto the glass substrate, it can modify the morphology of the cobalt oxide nanostructures. This is due to the elimination of the lattice mismatch between the nanostructures and the glass.

4. Conclusion

Nanostructures of Co_3O_4 have been synthesized using two step solution route technique. The effect of depositing a seed layer prior to the deposition of the secondary layer had been observed and was found to have a significant effect on the formation of various nanostructures; nanorods, nanowalls, nanoplates and nanofibrils. Though the temperature for the growth of these nanostructures is relatively low it was also observed that upon annealing it at 450°C for one hour could lead to the formation of Co_3O_4. This observation gives light to developing nanostructures at low temperature. Most importantly understanding the formation of various nanostructures by simply adding a seed layer could lead to synthesizing more complicated structures with unique or tailored properties for a particular application.

5. References

[1] Kuwabara T., Nishizawa B., Nakamura K., Ikeda Y., Yamaguchi T., Takahashi K.(2015). Electrocatalytic activity of electrodeposited cobalt oxide films to produce oxygen gas from water. *Journal of Electroanalytical Chemistry 740* (2015) 14–20.

[2] Nguyen T.T., Nguyen V.H., Deivasigamani R.K., Kharismadewi D., Iwai Y., Shim J.J. (2016). Facile synthesis of cobalt oxide/reduced graphene oxide composites for electrochemical capacitor and sensor applications. *Solid State Sciences* 53 (2016) 71-77.

[3] Zhang H., Liu S. (2017). A combined self-assembly and calcination method preparation of nanoparticles-assembled cobalt oxide nanosheets using graphene oxide as template and their application for non-enzymatic glucose biosensing. *Journal of Colloid and Interface Science 485* (2017) 159–166.

[4] Wang X., Yan C., Sumboja A., Lee P.S. (2013). High performance porous nickel cobalt oxide nanowires for asymmetric supercapacitor. *Nano Energy* (2014) 3, 119–126.

[5] Jagadale A.D., Kumbhar V.S., Lokhande C.D. (2013). Supercapacitive activities of potentiodynamically deposited nanoflakes of cobalt oxide (Co3O4) thin film electrode. *Journal of Colloid and Interface Science 406* (2013) 225–230.

[6] Pudukudy M., Yaakob Z., Narayanan B., Gopalakrishnan A., Tasirin S.M. (2013). Facile Synthesis of Bimodal Mesoporous Spinel Co_3O_4 Nanomaterials and their Structural Properties. *Superlattices and Microstructures 64 (2013) 15–26.*

[7] Kung C.W., Chen H.W., Lin C.Y., Vittal R., Ho K.C. (2012). Synthesis of Co3O4 nanosheets via electrodeposition followed by ozone treatment and their application to high-performance supercapacitors. *Journal of Power Sources* 214 (2012) 91-99.

[8] Kim K.H., Umakoshi T., Abe Y., Kawamura M., Kiba T. (2015). Determination of effective growth time for zinc oxide nanorods using chemical solution deposition. *Superlattices and Microstructures* 88 (2015) 150-153.

[9] Manigandan R., Giribabu K., Suresh R., Vijayalakshmib L., Stephen A., Narayanan V., (2013). Cobalt Oxide Nanoparticles: Characterization and its Electrocatalytic Activity towards Nitrobenzene. *Chemical Science Transactions 2013, 2(S1), S47-S50.*

[10] Patil D., Patil P., Subramanian V., Joy P.A., Potdar H.S. (2010). Highly sensitive and fast responding CO sensor based on Co3O4 nanorods. *Talanta* 81 (2010) 37–43.

[11] Pourshaban E., Abdizadeha H., Golobostanfarda M.R. (2015). ZnO Nanorods Array Synthesized by Chemical Bath Deposition: Effect of Seed Layer Sol Concentration. *Procedia Materials Science 11* (2015) 352 – 358.

[12] You H.C. (2014). Transistor characteristics of zinc oxide active layers at various zinc acetate dihydrate solution concentrations of zinc oxide thin-film. *Journal of Applied Research and Technology* 13 (2015) 291-296.

IOP Conf. Series: Materials Science and Engineering **205** (2017) 012029 doi:10.1088/1757-899X/205/1/012029

Synthesis and Characterization of Manganese Sponge doped with TiO_2

Mallikarjun patil

Department of Aerospace Engineering, Alliance University, Bangalore, India
mallumalluae@gmail.com

Abstract. The aviation and mechanical industry has seen tremendous growth in recent years. While these trends only continue, the materials on development of better fabrication technologies. Spongy materials are projected as favourable materials for the future as porous in nature. The current paper devotes to enhance mechanical properties by the use of Manganese sponge doped with TiO_2 possess an outstanding array of properties not readily achievable with other materials. Samples were synthesized by Kroll process and solid-state reaction. The specified materials compares well with the presently used aeromechanical materials in all terms. The scope for further research work in this area will be outlined.

1. Introduction

Now a days, the term "sponge material" is often used for any kind of metallic material contains voids, but generally sponge materials are produced in the partial solid state. According to fabrication process, sponges are basically materials who boundaries are made up of solid lattice, the interior are voids and individual cells are separated from each other by the metal. It is usually closely embedded by uneven pores in it, and for a sponge material porosity level generally be less than 70-75%. Sponge material have continuous network of metal pieces and followed by the co-existence network of the empty space. The certain properties of "material sponge", generally they exhibit like having a versatile potential in many industrial fields because of capability of producing light weight structures, high energy absorption and emission phenomenon, having low density and stiffness. As they are spongy in nature must exhibit the wide range of porosity and volume fraction in microstructure scale. According to the application requirement, the term porosity plays a very high command role in deciding the optimum shape and also mechanical properties of the structure. Manganese sponge is one of the best contender material to be used in manufacturing turbine blades because of having high strength to weight ratio, propellants tanks for various space vehicles as non-corrosive in nature etc.

Figure.1. Macro picture of (TiO_2 doped Manganese sponge)

IOP Conf. Series: Materials Science and Engineering **205** (2017) 012029 doi:10.1088/1757-899X/205/1/012029

2. Synthesis of Manganese sponge doped with TiO_2

Metallurgy for synthesizing sponge with TiO_2 is an advantageous task of increasing the possibility of obtaining the simplex geometry of the material. Initially, manganese chloride sintered at low temperature effectively with cobalt metal and ethanol as a binder at 473 K gives "manganese sponge" by Kroll process as the desired product followed by bi-products are hydrochloric acid and carbon dioxide. It has shown a good control over porous structures as follows:

$$3MnC + CO + 6C_2H_5OH \xrightarrow{473\ K} 3Mn + 2Co_2 + 6HCl + COO \qquad (1)$$

Getting manganese sponge is followed by doping TiO_2at 1260 k (sintered)

$$3Mn + TiO_2 \xrightarrow{1200\ K} \quad + TiO_4 \qquad (2)$$

The product reveals in equation (2) is the titanium dioxide [2] dopant in manganese sponge.

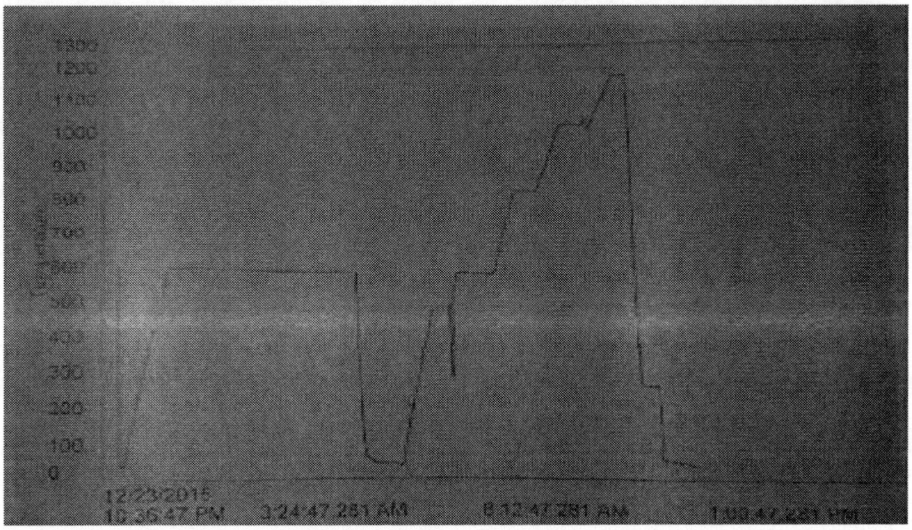

Figure.2. Sintered graph (TiO_2 doped Manganese sponge)

3. Characteristics of manganese sponge doped with TiO_2

The core product after doping the desired sponge with TiO_2 is 99.64%. Dopant is just about 0.24% in the overall stoichiometry. It is phenomenal that after doping with titanium oxide for the flexibility in the bit amount leads the manganese output closer to the currently glowing titanium sponge following basic properties compares well with the current titanium product.
The following table gives the some experimental results:

Table 1. Overall Properties and Characteristics of Manganese sponge doped with TiO_2 as well as Titanium sponge:

Properties	Manganese sponge Doped with TiO_2	Titanium sponge

IOP Conf. Series: Materials Science and Engineering **205** (2017) 012029 doi:10.1088/1757-899X/205/1/012029

Phase	Partially-solid	Semi-solid
Granule size (nm)	0.49	0.70
Boiling point (K)	2334	3533
Melting point (K)	1579	1941
Hardness (BHN)	122-153	141-160
Porosity (per mole)	4.40 %	4.13%
Thermal conductivity (W/m.K)	7.81	29.7
Density (gm/cm^3)	7.21	4.48
Strength to weight ratio (kN.m/kg)	277.58	287.7
Tensile strength (MPa)	1154.24	1300.23

3.1. SEM analysis of TiO_2 doped manganese sponge

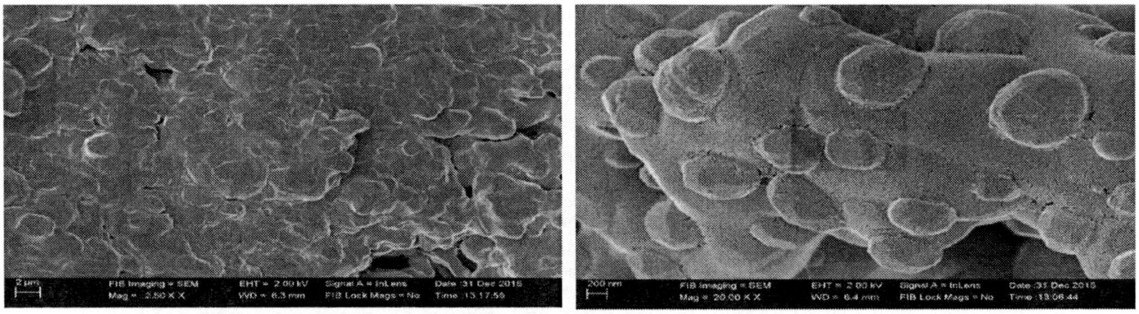

Figure.3. SEM micro pictures of (TiO_2 doped Manganese sponge)

- At 200 nm, it is clear as per the characterization of the material has been described about the granule size also the van der wall forces also helps in binding overall structure from the molecular level.
- At 1µm level although at per the magnification it is showing the certain trailer of the porosity originates from that point. As the magnification varies is being showing clearly about the binding force strengthens the material from micro level.
- At 200µm, the surface look smooth with some bit of holes, but the overall characteristics compares well with the current scenario of the light-weight materials.
- At the certain level of 10µm, the surface is showing high density, though it is affording healthy amount of porosity ratio in it.

4. Conclusion

The paper visits the synthesis and the various properties of Titanium dioxide doped manganese sponge. The vital intension to consider porous material for the manufacturing applications has been discussed, giving solid evidence. The paper intended to show the fabrication of manganese sponge at the initial stage and then pursued by doping it with titanium dioxide. The analysis shown above tries to provide quite a few reasons for replacing titanium products by the doped manganese compound. The properties and overall picture resembles with each other and therefore the new products can be seen comparing well soon with it in the aviation manufacturing field.

5. Uses

The research so far shows, manufacturing of airplane's wing, flap and compressor blade by using manganese sponge doped titanium dioxide results the overall **specific fuel consumption** by any twin-aisle engine would get better in keeping the idea implemented which shows a remarkable journey in the field of Aviation.

As per the survey, generally the entire cost of an airplane is somewhere around **$400 million**. As mentioned earlier as well, reduction in few percent can lead aviation industry to a cost effective manufacturing by implementing these ideas in certain prospects and could get a compensation of a few thousand dollars by implementing the above proposal with high overall efficiency. So, the research and development work on these couple of proposals are worthwhile and investment would be highly rewarding.

6. Acknowledgments

I am very grateful to parents, mentor Simranjeet Singh, Research Assistant, Department of Aerospace & Mechanical engineering at Alliance College of Engineering and Design along with the faculty members of Alliance University for their valuable guidance, motivation and technical support.

7. References

[1] Ziya ESEN, Elif TARHAN BOR, Sakir BOR, *Characterization of loose powder sintered porous titanium and Ti6A14V alloy,* Turkish j. Eng. Env. Sci.33 (2009), 207-219.c_T"UB'ITAK,doi:10.3906/muh-0906-41.

[2] M.V.de Oliveira, L. C. Persia, L .M. Reis, *PROCESSING AND CHARACTERIZATION OF TITANIUM SPONGES WITH DIFFERENT POROSITY LEVELS,* 17□ CBECIMAT (FOZ DO IGUACU/BRAZIL, NOVEMBER 15-19, 2006).

[3] Kimiyoshi *Ono, Kizashi Takahashi, Direct production process of high hardness titanium sponge, US 3359093 A, Dec 19 1967*

[4] I Sh Trakhtenberg, A B Borisov, V I Novozhonov, A P Rubshtein, A B Vladimirov, A V Osipenko, V A Mukhachev, E B MAKAROVA, *Mechanical properties and structure of porous titanium obtained by sintering compacted titanium sponge, Strength and Plasticity, The physics of metals and metallography,* January 2008, Volume 105, Issue 1, pp 92-97, first online: 22 February 2008.

[5] Robert A. Clyde, *Supported catalyst,* US 3998758 A, US 05/605,677, Dec21,1976 US 05/605,677, Dec 21,1976.

[6] Chisholm Douglas S, *Continuous production of titanium sponge,* US 3067025 A, Dec 4,1962.

IOP Conf. Series: Materials Science and Engineering **205** (2017) 012030 doi:10.1088/1757-899X/205/1/012030

The Preparation and Characterization of Natrolite Synthetized by Purified Attapulgite

H J Li [1,2,a], X D Zhou[1,b], J M Zhang[2,c], X Y Wu[3,d] and H B Gao[4,e]

[1] Faculty of Water Resources and Hydraulic Power, Xi'an University of Technology, Xi'an, 710048, Shaanxi, China
[2] School of Environment and Chemistry Engineering, Xi'an Polytechnic University, Xi'an, 710048, Shaanxi, China
[3]Shaanxi construction and Installation Group Co. LTD., Xi'an, 710043, Shaanxi, China
[4]China Construction Seventh Engineering Division. Corp.LTD, Xi'an, 710048, China
Email:[a]email:lhj861106@126.com,[b]email:zhouxd@mail.xaut.edu.cn,[c]email:jm0865@sina.com [d]1058893607@qq.com, [e] email: gaohongbo2011@qq.com

Abstract. This paper mainly researched the hydrothermal synthesis of Natrolite, using amorphous silicon source from the purified attapulgite. The effects of silicon source, silicon aluminum ratio, crystallization time and crystallization temperature on the synthesis of natrolite were investigated. The results showed that the optimal synthesis condition of natrolite was: Hydrothermal activated ATP with NaOH was silicon source, silicon aluminum ratio was 10:1, crystallization time lasted to 72h and crystallization temperature was 150℃, the template was removed by calcining 8 hours at 550℃. The structural formula of obtained natrolite is $Na_2Al_2Si_3O_{10}•2H_2O$.

1. Introduction

Zeolites, a family of crystalline microporous aluminosilicate materials,its micropores are defined as pores with diameters <2nm[1]. Due to its high thermal stability, hydrothermal stability and good shape-selective ability, Zeolites is widely used in adsorption, ion exchange and catalysis[2-4]. At present, there are many methods to synthetize the zeolite, such as hydrothermal synthesis, microwave synthesis, solid phase syntheses and so on. A wide variety of zeolite include MFI, Beta, MCM-41, Natrolite ,et al. The natrolite framework is composed of T_5O_{10}(T = Si, Al, Ga, Ge...) tetrahedral units, which in the case of natrolite have a SiO_4 tetrahedron as a central polyhedron and the surrounding two SiO_4 and two AlO_4 tetrahedra in alternation to satisfy the Loewenstein rule[5].

Yongjae Lee reviewed the effect of pressure on the natrolite structure[5].The synthesis of sodium natrolite is affected by various factors, such as crystallization temperature, crystallization time, the ratio of silicon to aluminum, and the kind of silicon source.Arezo Hatamifard[6] researched that the green synthesis way of a natrolite zeolite/palladium nanocomposite ,and its application as a reusable catalyst for the reduction of organic dyes in a very short time. Siavash Bahari[7] used natrolite zeolite as a natural and reusable catalyst and the results of the present work showed the desired product in excellent yield. SB Hong [8] invented a method for selectively isolating hydrogen or helium using a natrolite-based zeolite. G.A.Mamedova[9] researched the hydrothermal synthesis of natrolite-type zeolite in the natural halloysite-obsidian system.

Attapulgite, called the palygorskite or stripe stone, is a kind of layer silicate mineral materials with silicon, aluminum, magnesium, iron, et al. The typical chemical molecular formula is that

$Mg_5Si_8O_{20}(OH)_2(OH_2)_4{\cdot}4\ H_2O$[10-12]. The unique layer chain structure and high specific surface area, enable attapulgite to have size/shape selective adsorption and catalytic function, it is used in synthesis of nano-composite adsorbents and control ling the environmental pollutants at resently years[13]. The main components of purified attapulgite by chemical or physical methods are mainly silicon and oxygen, and also contain a small amount of metal ions. Molecular sieve is synthesized by purified attapulgite as silicon source, the effects of silicon source, the ratio of Si/Al under the different crystallization temperature and the synthesized time were discussed. The aim of this paper is that the preparation and characterization of natrolite synthetized by purified attapulgite.

2. Experiment and Characterization

2.1 The Instrument and Reagent

(1) Instrument: SXC-2-13 muffle furnace; DZF - 6030 vacuum drying oven;AL204 electronic balance; HJ-4a, heating magnetic stirrer. Thermal water kettle; Vacuum suction filter machine.ASAP-2020HD88 Accelerated Surface Area and Porosimetry System.

(2) Reagents:purified attapulgite,hydrochloric acid, alumina, sodium hydroxide, TEAOH (tetraethyl ammonium hydroxide) .

2.2 The acquisition of silicon source and purified attapulgite

(1)NaOH modified ATP: A certain amount of nature attapulgite and sodium hydroxide powder mixed evenly in the crucible; The sample in the crucible is calcined at 550℃ at a speed of 1℃/min for 2 hours in a muffle furnace, then grinds after cooling. The samples are mixed with NaCl 1mol/L solution, and adjusted pH to 6 with 1mol/L HCl, filtrating, drying and standbying.

(2)Acid thermal attapulgite: 2.00g of Alkali modified attapulgite was placed in 50 mL beaker, then a certain amount of 9% (wt) of hydrochloric acid was added. The mixture was stirred well with a glass rod, then the hydrothermal acidification in the reaction kettle and sealed with PTFE lining was carried out at 180°C for 12 hours. The resulting material was filtered and dried at 100 °C. The raw attapulgite contains a certain amount of silicon source and lots of metal impurity, but metal impurity was removed after purification, the mainly rest is SiO_2.

2.3 Hydrothermal synthesis of Natrolite

Weighing the quality of purified attapulgite as silicon source, the aluminum oxide, NaOH and TEAOH in a certain proportion. The above mixture placed in a beaker by use of magnetic stirring for One hour. Natrolite was synthesized by hydrothermal crystallization from the product in autoclave at 150 ℃ for 72h,followed by calcining at 550 ℃to remove the urea template.

2.4 Characterization

Rigaku D/max-Rapid Ⅱ X - ray diffraction (XRD) analysis was used to determine the crystal phase , equipped with Cu Kα source operated at 40 kV and 40 mA with a scan rate of 8°/min, a scan step of 0.02°,the scan range of 5°~70°. Scanning electron microscopy (SEM) analysis was performed on a Quanta-450-FEG electron micro-scope, fabricated by FEI of the UK. Nitrogen adsorption-desorption isotherms were obtained on an ASAP-2020HD88 Accelerated Surface Area and Porosimetry System.

3. Results and Discussion

3.1 NaOH modified ATP

By a certain mass ratio of ATP and NaOH mixed evenly into a muffle furnace at 550 ℃, roasting by the aforementioned method, the morphology and composition analysis by SEM and EDS, the results are as following.

Figure.1 shows that Na element is found in modified product with NaOH compared without Na ion in the original attapulgite, which indicates the composition of the substance from the skeleton is changed through roasting. In the air, the metal ion content decreased under the order: Fe > K > Mg > Al>Cu, K.

IOP Conf. Series: Materials Science and Engineering 205 (2017) 012030 doi:10.1088/1757-899X/205/1/012030

K ion as active metal elements in the framework of ATP is removed by the high temperature roasting, Cu is away from the framework because of the replaced of Na.

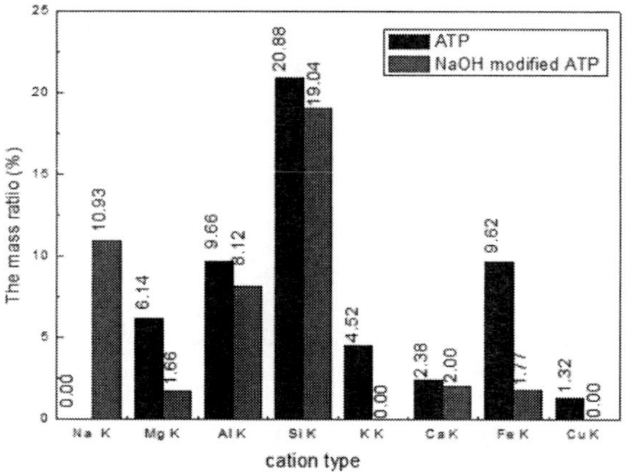

Figure.1 The composition analysis of modified product with NaOH

Figure. 2 SEM of modified product with NaOH in air atmosphere

From Figure.2, to observe the surface morphology of the product in 20.00kv, 4000 times, Calcined in air is granular and loosely dispersed, and the surface of a flocculent material existence.

3.2 Acid thermal the modified ATP with NaOH

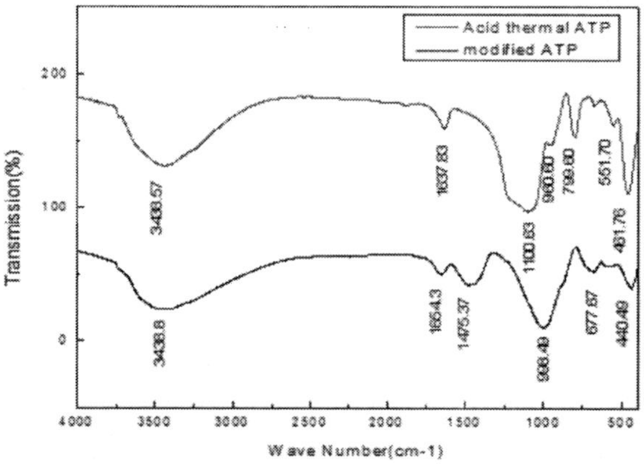

Figure.3 FT-IR analysis of activated products under hydrothermal conditions

Figure.4 SEM of Acid thermal the modified ATP with NaOH

Figure.3 shows: The peak of 960.60 cm^{-1} in the graph is Si-O asymmetric stretching vibration peak, in T-O tetrahedron, reflecting the $Si-O_4$ stretching vibration in I direction. It proved that the basis of modified ATP with HCl and NaOH is Si and O.

From Figure.4, to observe the surface morphology of products in 10.00kv, 60000 times, the extraction of silicon source is the size of nearly spherical material, and the surface is flocculent. It shows that the existence form of silicon is amorphous state.

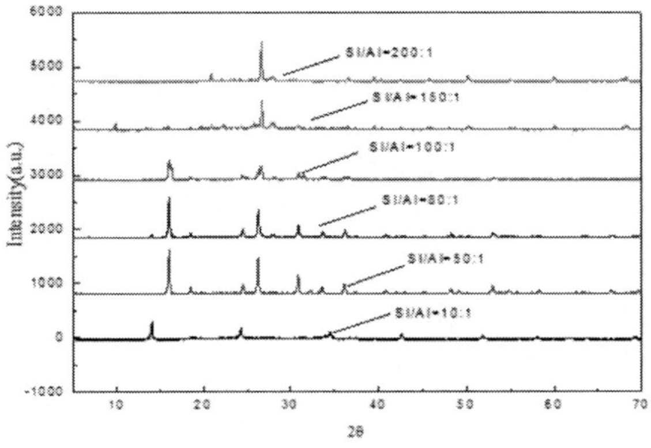

Figure.5 XRD analysis of syntheticproducts with different ratio of Si/Al

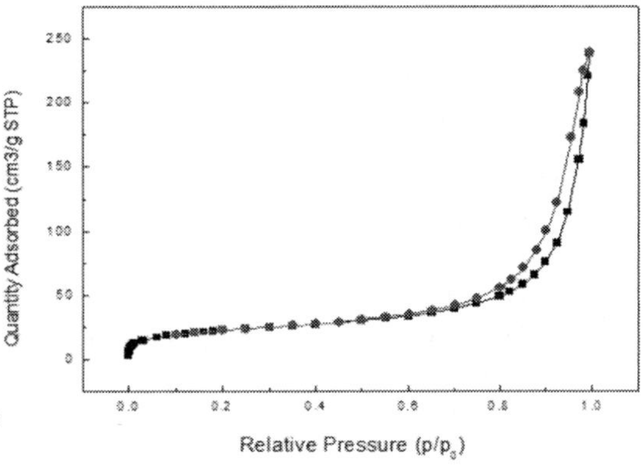

Figure.6 N_2 adsorption and desorption isotherm of Natrolite

3.3 XRD and BET analysis of Natrolite

The molecular sieve was synthesized by the above method, the structure and species were characterized by XRD as Figure.5.The XRD results: when the ratio of Si/Al is 10:1, the synthesized product is natrolite, the molecular formula is $Na_2Al_2Si_3O_{10}•2H_2O$. Strong diffraction peaks appear at 2θ=15.05, 25.72, and 43.55°, which corresponding belongs to Natrolite at (111), (440), (911), crystallographic plane.

The N_2 adsorption-desorption isotherm of Natrolite belongs to typical type-Ⅰ isotherm. Due to the strong interaction between the adsorbed material and the pore wall, the adsorption curve starts at a low relative pressure. The adsorption and desorption curves are basically the same. Single point BET specific surface area of Natrolite is 84.34 m^2/g, the average pore is 1.76nm.It belongs to microporous zeolite.

4. Conclusions

Modified ATP with NaOH in air atmosphere decrease Fe, Mg and Al while increasing the Na, K and Cu are completely removed .To purify more, the above product is hydrothermal active with HCl .The basis of Acid thermal ATP with NaOH is Si and O.XRD shows that hydrothermal activated ATP with NaOH as silicon source, Si/Al was 10:1, crystallization time was 72h and crystallization temperature was 150℃, the template was removed by calcining 8 hours at 550℃. The structural formula of obtained natrolite is $Na_2Al_2Si_3O_{10}•2H_2O$. Single point BET specific surface area of Natrolite is 84.34 m^2/g, the average pore is 2.76nm.It belongs to microporous zeolite.

5. Acknowledgements

This work was financially supported by the Shaanxi Provincial Department of education special scientific research project(15JK1327).

6. References

[1] Molecular sieve group, Dalian Institute of Chemical Physics, Chinese Academy of Sciences. Zeolite molecular sieve [M].Peking: Science Press, 1978.
[2] K.Suzuki, K.Ikari, H.Imai, Synthesis of silica nannoparticles having a well-ordered mesostructure using a double surfactant system, J.Am.Chem.Soc.[J].2004,126(2):462-463.
[3] L.H.Chen, X.Y.Li, J.C.Rooke, Y.H.Zhang, F.S.Xiao, B.L.Su, Hierachically structured zeolites: synthesis, mass transport properties and applications, J.Mater. Chen. [J].2012,22(34):17381-17403.
[4] D.Bonenfant,M.Kharoune,P.Niquette.Aduances in principal factors influencing carbon dioxide adsorption on zeolites , Sci.technol.Adv.Mater. [J].2008, 9(1):14-21.
[5] Yongjae Lee, Chi-Chang Kao and Thomas Vogt. J. Thermal Expansion of the Superhydrated

IOP Conf. Series: Materials Science and Engineering **205** (2017) 012030 doi:10.1088/1757-899X/205/1/012030

Small-Pore Zeolite Natrolite.Phys. Chem. C[J]. 2012, 116, 3286–3291.

[6] Arezo Hatamifard, Mahmoud Nasrollahzadeh and Janusz Lipkowskib, Green synthesis of a natrolite zeolite/palladium nanocomposite and its application as a reusable catalyst for the reduction of organic dyes in a very short time.RSC Adv. [J].2015,5, 91372-91381.

[7] S Bahari，SM Sajadi. Natrolite zeolite: A natural and reusable catalyst for one-pot synthesis of α-aminophosphonates under solvent-free conditions.Arabian Journal of Chemistry[J]. 2012, 32.

[8] SB Hong, Method for selectively isolating hydrogen or helium using a natrolite-based zeolite ,and nvel natrolite-based zeolite .EP2394729A1[J].2011,12.

[9] GA Mamedova, Glass Physics and Chemistry[M].2014, 40(3):380-383.

[10] P.F.Wang, A.Lv, J.Hu, The synthesis of SAPO-34 with mixed template and its catalytic performance for methanol to olefins reaction,Microporous and Mesoporous Material[J].2012,152 (1):178-184.

[11] A.Vinu, T.Mori, K.Ariga. New families of mesoporous materials, Science and Technology of Advanced Materials[J].2006,7(8):753-771.

[12] L.H.Chen, X.Y.Li, J.C.Rooke. Hierarchically structured zeolites: synthesis, mass transport properties and applications.Journal of Materials Chemistry[J].2012,22（34）:17381-17403.

[13] M.S.Holm, E.Taarning, K.Egeblad. Catalysis with hierarchical zeolite, Catalysis Today, 2011, 168(1):3-16.

IOP Conf. Series: Materials Science and Engineering **205** (2017) 012031 doi:10.1088/1757-899X/205/1/012031

Analysis of Wood Structure Connections Using Cylindrical Steel and Carbon Fiber Dowel Pins

Mikhail A. Vodiannikov[1*], Dr. Galina G. Kashevarova[1]

[1]Perm National Research Polytechnic University, Perm city, Russia, Komomilski Prospect, 29. Russian.

email: vodyannikov@mail.ru

Abstract. In this paper, the results of the statistical analysis of corrosion processes and moisture saturation of glued laminated timber structures and their joints in corrosive environment are shown. This paper includes calculation results for dowel connections of wood structures using steel and carbon fiber reinforced plastic cylindrical dowel pins in accordance with applicable regulatory documents by means of finite element analysis in ANSYS software, as well as experimental findings. Dependence diagrams are shown; comparative analysis of the results obtained is conducted.

1. Introduction

Wood is one of the most ancient building materials used up to the present days. The use of structural engineered wood products is undermined by its chemical, aesthetic, environmental, economic, etc. features. The items made of solid wood (timber, board, semi rigidly connected composite members), as well as glued laminated wood structures (glulam) are used in modern construction.

Nowadays laminated wood products are commonly used in many countries for wide span structures of different architectural shapes and sizes. An example of commercial structures is arch and noggin systems used for the construction of warehouses, airplane and ship sheds, bridges, etc. [1]. In civil construction, glulam is used for roofing of public buildings, swimming pools, sport and agricultural facilities [2].

2. Observation results of corrosive environment on materials

Based on long-term observation experience, wood is less subject to destructive environmental impacts. Wood performance properties in corrosive environment are advantageous in comparison to steel and reinforced concrete. During the period of service, no emergency situations due to wood's chemical corrosion occurred. However, wood structure connections designed nowadays with nails, steel bolts, dowels, split elements are subject to corrosion, which is the most common cause for emergency situations. Further, threats influencing the strength of structures manufactured of timber and steel dowel pins are considered.

The most significant wood strength reducing factor is humidity. By means of experiment we have determined that wood moisture content does not significantly depend on ambient conditions. Microclimate inside the structure is mostly constant with insignificant fluctuations depending on season. Structural wood moisture content accumulation and penetration into the section depends of the service period. Figure 1 shows the penetration of moisture content into structural element section with thickness over 150 mm and 40 years of service period (the data is obtained empirically by selection and analysis of samples of existing structures).

IOP Conf. Series: Materials Science and Engineering **205** (2017) 012031 doi:10.1088/1757-899X/205/1/012031

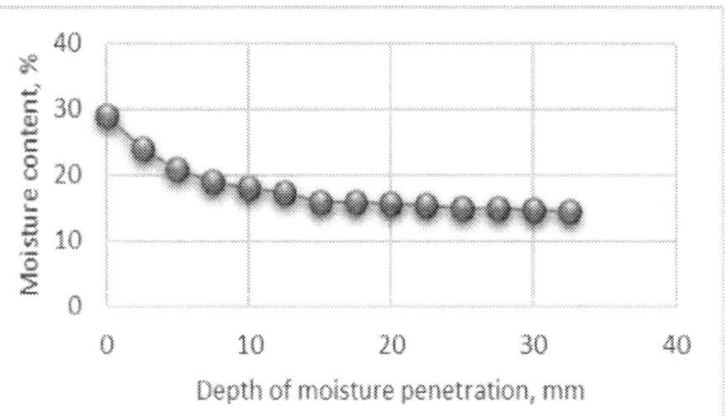

Figure 1. Moisture distribution throughout section

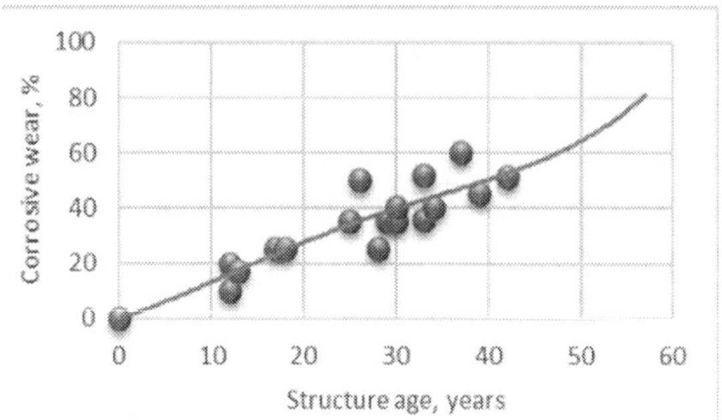

Figure 2. Dependence of corrosive wear of steel dowel pins and structure age

The diagram shows that the process of moisture content penetration is most intense during first 5…8 years. After that, the process slows down and the depth of moistened part does not exceed 1/3 of element height. The empirical evidence has shown that beginning from 20 years the partial destruction of wood section does not lead to avalanche-type destruction through its whole thickness; instead, stress redistribution in the section occurs. Experimental evidence and design values let us predict that the service life of bearing wood structures subject to repetitive moisturizing would be not less than 60 years.

In order to make a comparison, we should consider corrosion test results for steel dowel pins used as anchor support structures for wood half-arches (Figure 2) of muriate potash warehouse. As shown in the Diagram, cross-section of steel cylinder dowel pins would be completely destroyed by corrosion in approximately 60 years of service in these conditions. Moreover, the bearing capacity will decrease in much shorter period.

One of reasonable solutions for the design of reliable joints of wood structures in corrosive environment is the replacement of steel elements with materials that are not affected by corrosion [3]. The most promising material of those is structural carbon fiber reinforced plastic.

Carbon fiber reinforced plastic (CFRP) includes all composite materials which are reinforced by carbon (graphite) fiber. Different types of plastics may be used. Carbon fiber is structured organic material which consists of 92-99% of carbon and is characterized by high modulus of elasticity, strength and lower unit weight. CFRP is used for building new bearing structures and their connections, as well as for structural reinforcement, repair and protection of existing facilities.

In wood structures, nowadays, CFRP is only used for section reinforcement by methods of the insertion of carbon rods into a heavy section at different angles, installation of CFRP layers between boards [4]; and as external armor layer.

In this study the possibility of the replacement of steel dowel pins (steel S255 [5]) with carbon fiber dowels (CFRP "MONSTEROD") in wood structure connections is analyzed.

Dowel connections are flexible connections which bend and prevent reciprocal displacement of the connected elements. In such connections, the force is distributed between the elements in contact. The samples are destroyed either because of the crushing of the connected elements, or due to the bending of dowels.

The typical performance of dowels that are subject to bending and the type of surrounding wood crushing were considered in cases of steel and carbon fiber dowel pins use. Full-scale experiments, bearing capacity estimation by engineering method as per National Regulations [4] and calculations in ANSYS software were conducted.

3. Full-scale experiments for defining bearing load capacity of dowel connection

The diagram of samples of dowel connection of wood structure and the applied loads are shown in Figure 3. The layout of dowels is performed in accordance with the requirements of cl. 7.18 [4].

The tests were conducted until sample destruction by incremental loading using compression-testing machine PGM-500MG4 with load limit up to 500 kN. The value of breaking load was taken from

testing machine dynamometer at the moment when continuous increase of the displacement of the loaded central connected element was still observed, although the load remained the same.
In order to measure the displacement, dial gauges were used. The moisture content of the samples amounted to 9.2 %. The temperature inside the room is 21.3°C, and the humidity is 15.7%.
Based on the data obtained from the testing of each material of dowel pins (steel and carbon fiber), load-displacement diagrams were performed (Figure 4).

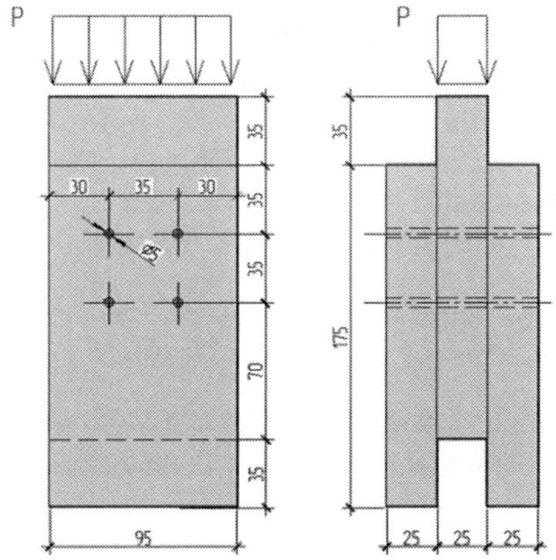

Figure 3. Sample loading diagram

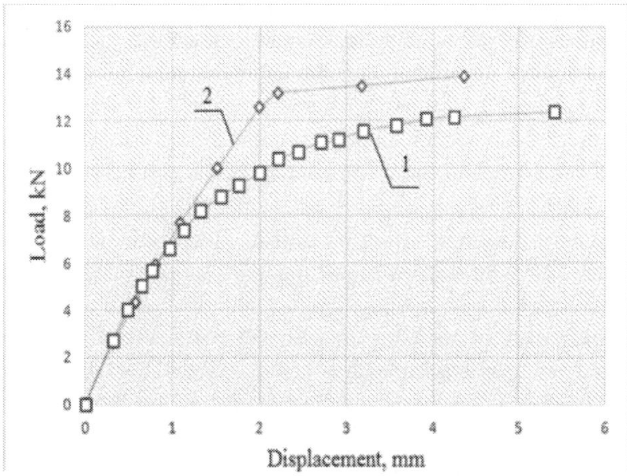

Figure 4. Load-displacement dependence for steel (1) and carbon fiber (2) dowel pins as per test results

The destruction of connections with steel dowel pins consisted of dowel bending and center wood element crack (Figure 5a). Breaking load was 11.1 kN, upon its reaching, the process of plastic flow of the connection began. At the same time, significant increase in displacement due to minor increase of loading was observed. A breakdown (a crack with a shatter in center element) took place when the applied load was 12.7 kN.
As for carbon fiber dowel pins, minor bending of dowel pins with insignificant polymeric matrix checking and opening of the samples at boards' planes (Figure 5b) were observed. However, no cracking of wood elements occurred. Breaking load was 12.6 kN, upon its reaching the deformation continued without further increase of load. Sample breakdown took place when the load was 13.9 kN, the boards moved to sides forming gaps between center and side boards.

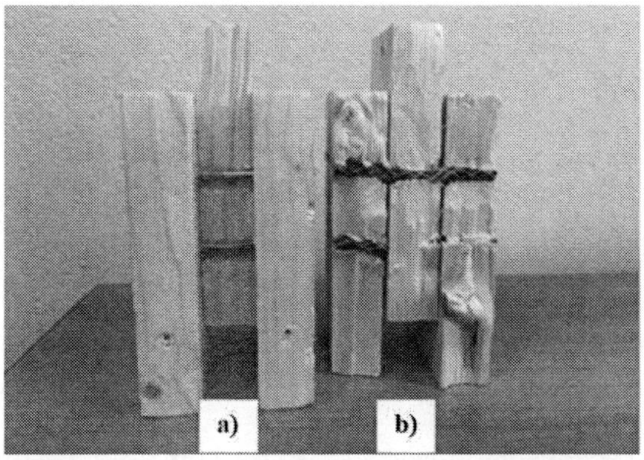

Figure 5. Type of sample destruction: a) for steel dowel pins; b) for carbon fiber dowel pins

4. Dowel connection calculation in accordance with national rules "Timber structures"
Design bearing capacity of a connection is determined from the following formula:
$T=T_{min}\cdot n_d\cdot n_s$, where

T_{min} – minimum bearing capacity of a dowel per one edge joint (conditional cross-section),
n_d – the quantity of dowels in the connection,
n_s – the quantity of cross-sections of one dowel.
The bearing capacity of one dowel per one edge joint Tmin is determined as the least of one of the following conditions [4]:
– wood crush in side element, $T_{cra}=0.8\cdot a\cdot d$
– wood crush in center element, $T_{crc}=0.5\cdot c\cdot d$
– dowel bending $T_{dowbend}=1.8\cdot d^2+0.02\cdot a^2$.
In these formulas, d is a dowel diameter; a and c are thicknesses of side and center elements respectively. Steel and carbon fiber dowels have same dimensions: $d = 5\ mm,\ a = c = 25\ mm$. These formulas are recommended to be used for steel cylindrical dowel pins. There is no regulatory documentation for carbon fiber dowels, therefore the same calculation formulas were used for them but also taking into account that CFRP has a better tensile strength than steel [7].
The design bearing capacity determined from this formula equals $Tmin = 5\ kN$ for the condition of center wood element bearing failure. Normally actual breaking load exceeds design breaking load 2 to 3 times [6]. The design breaking load taken is $Nu = 12.5\ kN$.

5. Computer modelling and dowel connection calculation

In order to analyze the strain-stress state of the dowel connection and verify the obtained experimental evidence we have created a model (for wood-steel and wood-carbon samples) in ANSYS software [8]. Considering structural symmetry and for the clarity of presentation of the middle part of the connection, the calculation is performed for a half of the section.
Because the calculation should consider friction forces at planes of wood-steel and wood-carbon contact, special contact elements were added to the model. Nonlinear contact problem "surface-surface" was to be solved. For the modelling of interface, the following elements were used: CONTACT173 (for wood surface) and TARGET170 (for dowel surface). The program uses contact elements for tracking of relative position of the two surfaces.
During the computer modelling, similarly to full-scale experiments, the loading from the machine to the sample was applied in gradual manner (increasing by 1 kN). Breaking load was defined for each sample separately.
Upon the completion of the test, the following results were obtained. For steel sample, the breaking load is in the range 14…15 kN. The reciprocal displacement of the nodes of wood elements located in several parts of the sample is 1.5 mm. Displacement isofields in wood are shown in Figure 6a. Load isofields and deformation type of steel dowel pins are shown in Figure 6b.

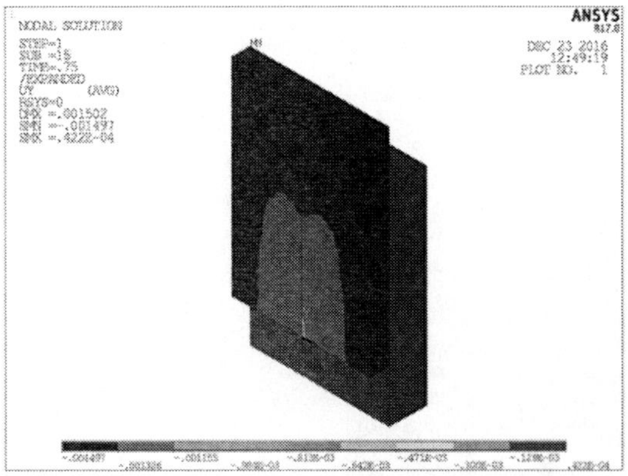

Figure 6a. Isofields of wood displacement

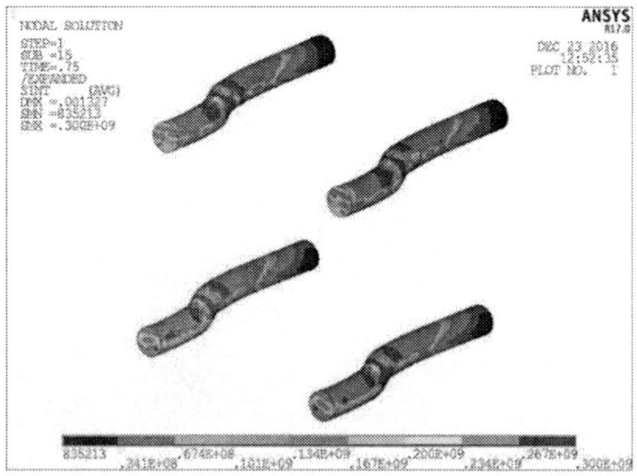

Figure 6b. Isofields of forces in steel dowel pins

Similar calculation was performed for carbon fiber dowels. The following results were obtained: breaking load amounted to 15…16 kN, reciprocal displacement of the nodes of wood elements located in several parts of the sample is 2.46 mm. Displacement isofields in wood are shown in Figure 7a. Load isofields and deformation type of carbon fiber dowel pins are shown in Figure 7b.

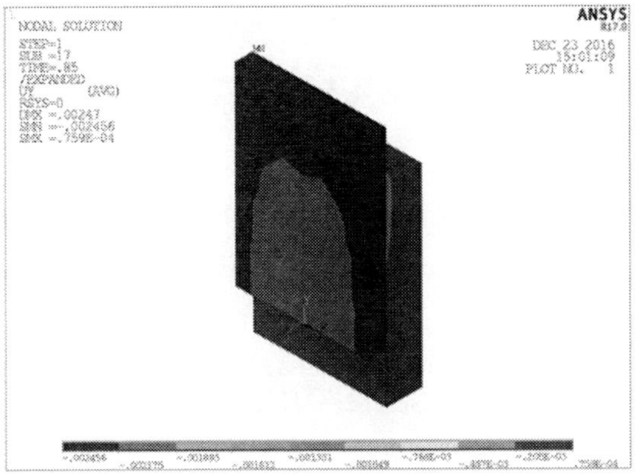

Figure 7a. Isofields of wood displacement

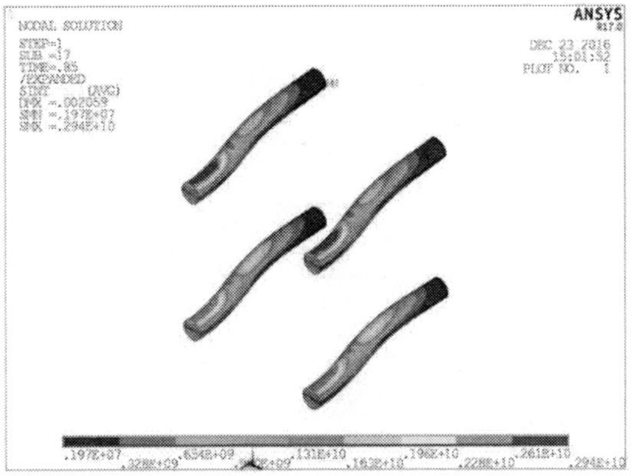

Figure 7b. Isofields of forces in carbon fiber dowels

6. Conclusion

The results of the analysis are as follows. Theoretical data, experiment evidences and the data obtained using ANSYS software verify each other. The spread in values is caused by wood structure imperfections (flaws), as well as lack of construction standards for the design of carbon fiber reinforced structures. As it was expected, the destruction of structures was caused by wood deformation. However, the type of dowel rod deformation is important for the connection bearing capacity definition and influences the type of sample destruction.

As it is seen from experimental evidence and theoretical data, in corrosive environment, the use of composite materials that are not subject to corrosion is determined by strong reasons. The use of carbon fiber reinforced plastic together with wood may be also determined by the similarity of their fibrous structure. The modern use of composites in aerospace industries has delivered their undisputable advantages compared to other materials. It is necessary to develop the knowledge of these types of materials, create regulatory and material and technical basis for their wider application.

7. References

[1] Kalugin A 2008 *Wooden constructions: the university tutorial* (Moscow: ASV).

[2] Lennartz M W 2016 *New Architecture in Wood: Forms and Structures* (Basel) vol 1

[3] Gugutsidze G and Draskovic F 2010 *Reinforcement of Timber Beams With Carbon Fibers Plastics (Slovak Journal of Civil Engineering)* pp 1-6

[4] SP 64.13330.2011 2011 *National construction regulations «Wood constructions»* (Moscow)

[5] SP 16.13330.2011 2011 *National construction regulations «Steel constructions»* (Moscow)

[6] Gappoev M 2004 *Wooden and plastic constructions* (Moscow)

[7] American Chemical Society National Historic Chemical Landmarks 2003 *High Perfomance Carbon Fibers* (Parma, Ohio)

[8] Basov K A 2002 *ANSYS examples* (Moscow)

IOP Conf. Series: Materials Science and Engineering **205** (2017) 012032 doi:10.1088/1757-899X/205/1/012032

Applying a New Parallelized Version of PSO Algorithm for Electrical Power Transmission

M Zemzami[1,3], **A Makhloufi**[2], **N Elhami**[3], **A Elhami**[2], **M Itmi**[1] **and N Hmina**[3]

[1]LITIS, Normandy University, National Institute of Applied Sciences, Rouen, France
[2]LOFIMS, Normandy University, National Institute of Applied Sciences, Rouen, France
[3]LGS, Ibn Tofail University, National School of Applied Sciences, Kenitra, Morocco
maria.zemzami@gmail.com

Abstract. In this paper, the optimization of an electric power transmission material is presented giving specific consideration on material configuration and characteristics. The nature of electric power transmission networks makes it hard to manage. Thus, giving need for optimization. So the problem of optimization of electric power transmission as considered in this paper is improving the performance and reliability of the electricity pylon; the objective is to maximize resistance to load while reducing material usage and cost. For this purpose, we suggest a new version of PSO algorithm that allows the amelioration of its performance by introducing its parallelization associated to the concept of evolutionary neighborhoods. According to the experimental results, the proposed method is effective and outperforms basic PSO in terms of solution quality, accuracy, constraint handling, and time consuming.

1. Introduction

Particle swarm optimization (briefed as PSO) is a nature-inspired algorithm that has shown outstanding performance in solving many realistic problems. It is now one of the most commonly used optimization techniques. However, premature convergence and computational costs are still a great constraints for PSO, as well as for all other metaheuristics, especially in complex optimization problems with time consuming objective functions.

In the standard PSO, all particles are directly updated by their offspring no matter whether they are improved. If a particle moves to a better position, it can be replaced by the updated. However if it moves to a worse position, it is still replaced by its offspring. In fact, the most particles fly to worse positions for most cases, therefore the whole swarm will converge to local optima. The exploitation and exploration of the search space represent two conflicting behaviors, which work together for solving the problem, and the right balance between them is integral to the performance of the PSO algorithm. The correlation of these behaviors with our approach is apparent, and this has been the inspiration behind the first part of our algorithm. The second part consists on minimizing computational costs by using parallel computation.

In order to test the proposed PSO algorithm, we designed and applied the developed optimization procedure to electricity pylon example. The application problem studied is weight minimization of bars truss, by finding optimal cross-sectional areas of the truss members. The results obtained solving this optimization problem using the proposed PSO algorithm is then compared with the results from ANSYS first order conventional optimization technique.

The present paper has sex sections including the introduction. In the next section, a brief description the basic PSO is given. In section 3 our proposed approach based on PSO algorithm is discussed. In Section 4 a description of the problematic is given. Section 5 describes the numerical results. The chapter finally concludes with Section 6.

2. Overview of PSO

2.1. Optimization

Optimization plays an important role in various disciplines such as Engineering Design, Manufacturing System, Computer Science, Economics, Artificial Intelligence, Operational Research and related fields. It is the process of trying to find the best, or optimal, solution to an optimization problem within a reasonable amount of time. An optimization problem is basically defined as: finding values of the variables that minimize or maximize the objective function while satisfying the constraints. The Optimization problems are centered on three factors: (1) an objective function which is to be minimized or maximized. (2) A set of unknowns or variables that affect the objective function. (3) A set of constraints that allow the unknowns to take on certain values but exclude others.

In order to solve optimization problems, optimization algorithms are used. The classification of optimization algorithm can be carried out in many ways. A simple way is looking at the nature of the algorithms, and this divides the algorithms into two categories: deterministic algorithm, and stochastic algorithms [1]. Deterministic algorithms are those whose behavior can be completely predicted as they are presented with same set of input and algorithm does same computations and produce same set results every time.

For stochastic algorithms, in general we have two types: heuristic and metaheuristic. Heuristic algorithms find solutions in a reasonable amount of time but there is no guarantee that optimal solutions are reached. On the other hand metaheuristic uses certain tradeoff a randomization and local search, as randomization provides a good way to move away from local search to the search on global scale which means metaheuristic algorithms intend to be suitable for global optimization. Nature-inspired metaheuristic algorithms are becoming powerful in solving modern global optimization problems.

Many modern metaheuristic algorithms were developed based on swarm intelligence in nature like PSO, this algorithm is based on the natural behavior and movement of insects, birds and fish.

2.2. Basic PSO

PSO is a population-based stochastic approach for solving continuous and discrete optimization problems [2]. As described by the inventers James Kennedy and Russell Eberhart in 1995 [2], "particle swarm algorithm imitates human (or insects) social behavior. Individuals interact with one another while learning from their own experience, and gradually the population members move into better regions of the problem space".

2.2.1. Formulation

Each particle is a point in the N-dimensional search space. A particle is represented by its current position Xi = (xi1, xi2,...,xij), and its current velocity Vi =(vi1 ,vi2 ,...,vij) i =1,2,...,q , j =1,2,...,n , where q is the swarm size. PSO tries to find the optimal solution to the problem by moving the particles and evaluating the fitness of the new position. Each particle searches for better positions in the search space by changing its velocity and position with every iteration, updating velocity depends on three factors; the particle's personal best position, the swarm's best position and the particle's previous velocity. The velocity and position of particles are calculated respectively as follows (1) and (2):

$$V(t+1)=V(t)+C1r1(Pb(t)-X(t))+C2r2(Pg(t)-X(t)) \tag{1}$$

$$X(t+1)=X(t)+V(t+1) \tag{2}$$

D is the dimension of the problem, i.e. the number of parameters of the function being optimized,
C1 and C2 are acceleration coefficients, i.e. cognitive and social parameters,
r1 and r2 are random numbers between 0 and 1,
XiD(t) is the position of particle i at time step t,
ViD(t) is the velocity of particle i at time step t,
PbD(t) is the best "remembered" individual particle position, at time step t,
PgD(t) is the best "remembered" swarm position, at time step t,
PSO has been the focus of many research efforts. Many improvements have been suggested and many more are still possible. Even in its current state, PSO is considered a fast and robust method for optimization of continuous nonlinear functions. In a PSO system, particles change their positions by

flying around in a multidimensional search space until reaching the convergence criteria. Each particle adjusts its travelling speed dynamically corresponding to the flying experiences of itself and its colleagues [3].

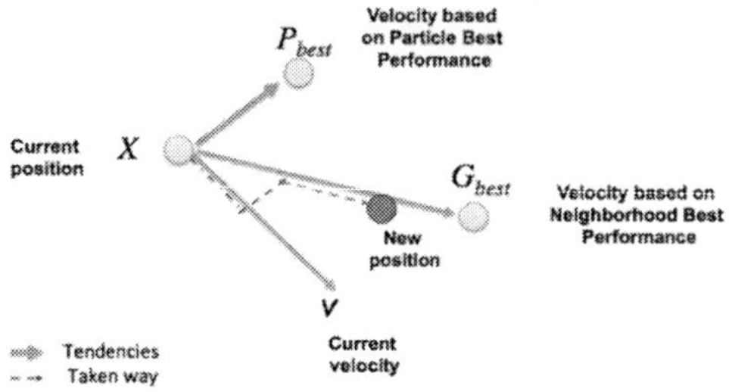

Figure 1. The movement of the particle by basic PSO

2.2.2. Algorithm

The basic Particle Swarm Optimization algorithm is a stochastic process proposed by [2] and it can be summarized as follows: (1) Initialize a population of particles: positions and velocities. The positions of the particles must be randomly distributed in the updating parameter space. (2) Calculate the velocity for each particle in the swarm using equation 1. (3) Update the position of each particle using equation 2. (4) Repeat steps 2 and 3 until convergence (met the stopping criterion).

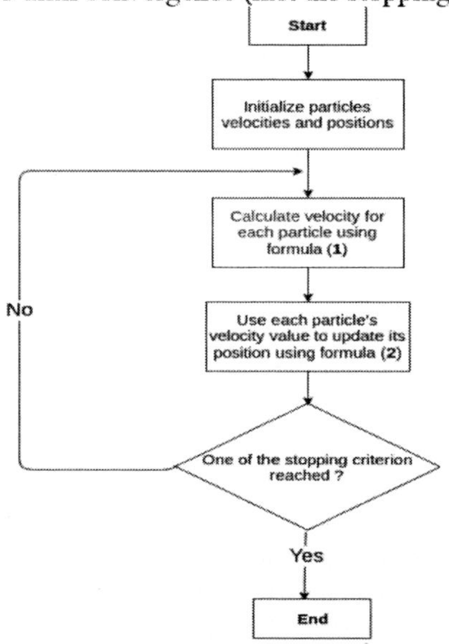

Figure 2. Basic PSO algorithm flow chart

2.2.3. Parameters

PSO has a number of parameters that determine not only its behavior but also its effectiveness. The establishment of PSO parameters affects greatly the PSO performance. Many specialists have already dealt with the improvement of PSO performance. From these studies, much effort has been invested to obtain a better understanding of the convergence properties of PSO. These studies based mostly on a better understanding of the basic PSO control parameters, namely the acceleration coefficients, inertia weight, velocity clamping, and swarm size [4] [5]. From these studies it can be concluded that the PSO is sensitive to control parameter choices, specifically the inertia weight, acceleration coefficients and velocity clamping. The wrong initialization of these parameters may lead to divergent or cyclic behavior.

In the initial version of the PSO, the values for C1, C2, the swarm size, the number of iterations, the dimension of the problem, and the stopping criteria; have to be selected. This selection has an impact on the convergence speed and the ability of the algorithm to find the optimum. A lot of variations that have been studied on the basic PSO equations to select a combination of values work well in a wide range of problems.

2.3. Neighborhood

The neighborhood constitutes the social network structure. It represents a particle with which each particle is able to communicate. There are two main types of neighborhoods:

The social neighborhood: this type of neighborhood represents the social proximity, neighborhoods are no longer the expression of the distance but the expression of the exchange of information, the neighbors are defined at the initialization and are not changed by the following. Once the network of social connections established, there is no need to refresh. It is a static neighborhood.

The geographical neighborhood: this type of neighborhood represents the geographical proximity, it is the most natural concept for the neighborhood particle swarm, neighbors are considered the closest particles. However, at each iteration, the new neighbors must be recalculated from a preset distance in the search space. It is a dynamic neighborhood that should be established and updated in each iteration. This is the kind of neighborhood that was retained in our approach.

The change of velocity of formula (1) is performed using a new term in the equation. It was introduced by [6], his picture appears in Figure 3, [7].

$$V(t+1)=V(t)+C1r1(Pb(t)-X(t))+C2r2(Pg(t)-X(t))+C3r3(Pn(t)-X(t))$$

With: Pn: the best position of the neighborhood; C3: the acceleration coefficient, also called social setting; r3: random number drawn from the interval [0,1].

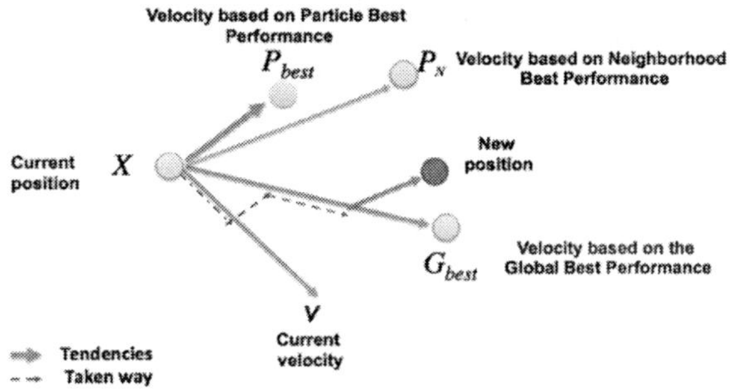

Figure 3. The movement of the particle by PSO Modified

3. The proposed approach

Nowadays engineering optimization problems often impose large computational demands, resulting in long solution times even on a modern high-end processor. The performance of serial PSO is affected when complex engineering optimizations are considered; this motivated the development of parallel optimization.

In the implementation of the classical PSO algorithm, all computations are done sequentially, that is where we get the idea of parallelization to improve the performance of the algorithm. Several authors have performed analysis to prove the performance of parallel PSO [8]-[12], the one we have adopted for our implementation, allows parallelizing computations. In fact processes are to perform calculations on set of particles located in different neighborhoods.

The threads, a sort of processes are executed in parallel for each iteration of the algorithm. Each thread executes the processing of an iteration of its particles' set, and waits for the other threads to finish their processing in order to update the neighborhoods and start a new iteration. This scenario repeats itself until a satisfactory solution is obtained: "achievement of stopping criterion". Our neighborhoods have the form of spheres, which are updated at each iteration: their centers evolve and the radius changes according to conditions relating to the number of neighborhoods.

The particularity of the approach consists to take advantage of the robustness of the PSO algorithm in choosing the right parameter setting, particularly the concept of dynamic neighborhood, in order to create diversity in research and in the sharing of information for a more optimal convergence. Besides the parallel computation that accelerates the calculations in order to have an optimal solution in an optimized computation time. The Figure 4 is a flow chart of the suggested approach. The reader is referred to [13] for a more details about this model.

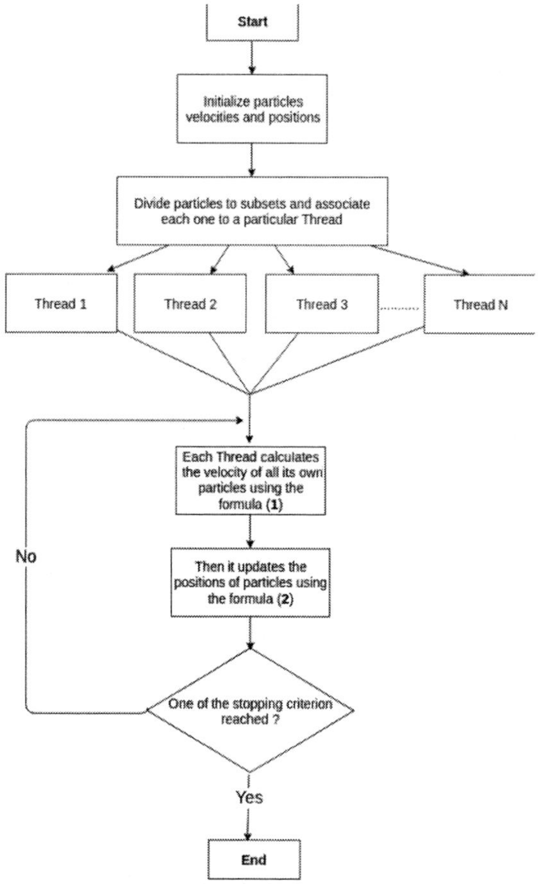

Figure 4. Flow chart of the proposed approach

4. Applying our proposed approach for electrical power transmission

In order to test the proposed PSO algorithm, we designed and applied the developed optimization procedure to electricity pylon example. The application problem studied is weight minimization of bars truss, by finding optimal cross-sectional areas of the truss members. The results obtained solving this optimization problem using the proposed PSO algorithm is then compared with the results from ANSYS first order conventional optimization technique. The static scheme of the problem is presented in Figure 5. Node positions and bars connections are fixed, only the sectional areas being the subject of the optimization. A pylon has a force of 18KN applied as shown below.

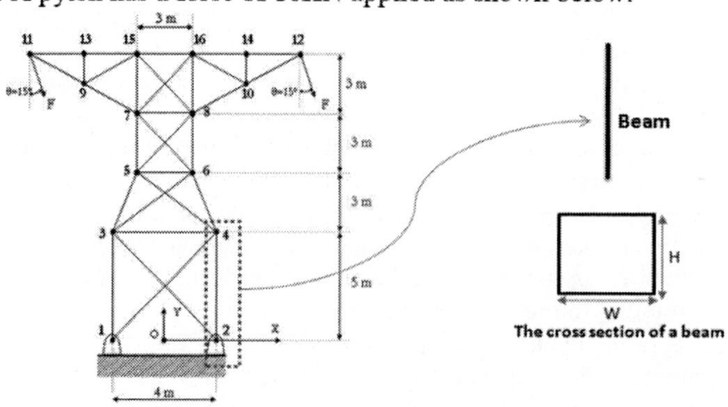

Figure 5. Geometric representation of the electricity pylon

The objective of this problem is to design a bars Steel truss structure of minimum weight (volume) that can with stand applied loads within the limits of allowable displacement. The material properties are Young's modulus EX = 200 GPa, Poisson's ration PRXY = 0.3 and weight density ρ of 7500kg/m3. The lengths of the truss members are fixed. The design variables or the input variables for the optimization problem are the cross sectional areas of the truss members. The cross sectional areas of the truss members are allowed to vary between 5 and 100 mm². The displacement limits is 9 mm in all direction. The geometry of the problem is shown in Figure 5. The structure is hinged at nodes (1, 2). ANSYS LINK1 element is used to model and simulate the structure for analyses [14]. The truss members are divided into four element groups as shown in Figure 6.

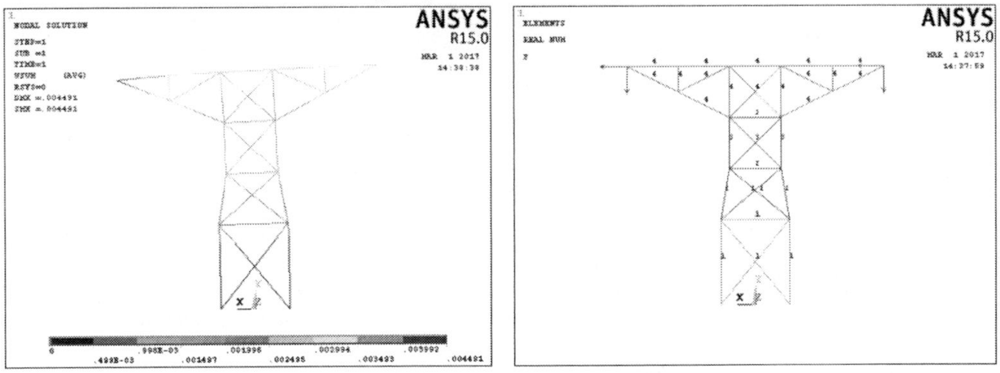

Figure 6. Element groups for the pylon

The four element groups signify four design variables. The design space in optimization can be considered as a four dimensional space where each design variable (cross-sectional area) represents four coordinates of the design space. The mathematical model can be expressed as:

$$A = [A_1, A_2, A_3, A_4]^T$$

$$f = \min W = \sum_{1}^{4} \rho_i \, . A_i \, . L_i$$

Subject to: $$g(A) = umax - uc \leq 0$$

$$A_{min} \leq A \leq A_{max}$$

Where $A = [A_1, A_2, A_3, A_4]^T$ are design variables for the cross section, f is the objective function. W is the total weight of the structure, ρ_i, A_i and L_i were the density, section area, bar length of the ith group bars, respectively. g(A) is the displacement constraints, umax et uc are the maximum displacement and the displacement limit of the ith group bars under various conditions. Amin and Amax are the minimum and maximum section size, respectively.

5. Numerical Results

The results obtained solving these optimization problems using the proposed PSO algorithm is then compared with the results from ANSYS first order conventional optimization technique.

The ANSYS first order optimization is a conventional optimization method where the true functions (Objective and Constraints) are used for the optimization. An APDL file that defines the pre-processing, solution, post-processing and the optimization phase is used for the conventional ANSYS optimization. In the optimization phase, the algorithm for optimization, the number of iterations ,the upper and lower bounds of the design variables and the limits on constraints are specified. The process converges within the specified iteration if a minimum value of objective function has been found that obeys the given constraints. If it does not converge a second iteration has to be setup with a new starting point. This goes on until a convergence is met within the specified constraints.

The same problem was also solved using the ANSYS First order optimization method. The maximum number of iterations was set as 20 for the ANSYS First order optimization process. The optimization problem converged to a minimum value of objective function given by W = 1.46E7 Kg. The optimal design variables were A1 = 26.86 mm², A2 = 25.95 mm², A3 = 24.86 mm² and A4= 24.54 mm². The convergence plot for the ANSYS First order optimization and the optimum values of the design variables are shown in Figure 7.

IOP Conf. Series: Materials Science and Engineering **205** (2017) 012032 doi:10.1088/1757-899X/205/1/012032

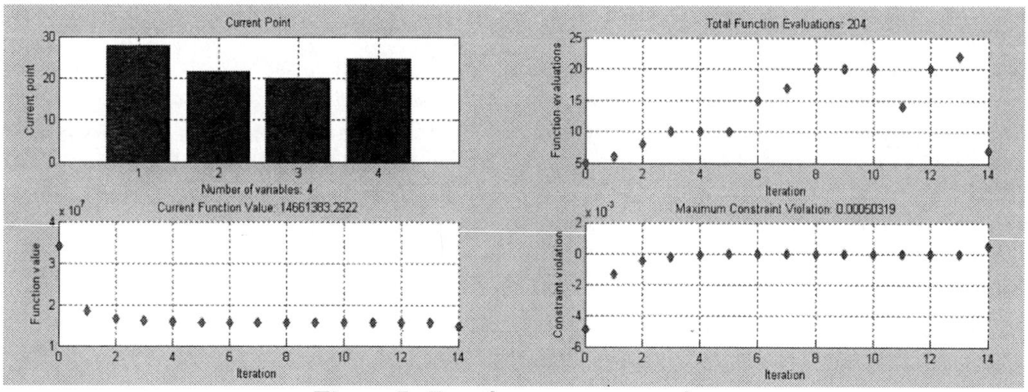

Figure 7. Function convergence plot

For the PSO algorithm, the procedure for optimization can be summarized in the following steps.

- Read the ANSYS APDL.
- Create response surface using second order polynomial based regression analysis using the candidate designs and the true responses. The generated response surfaces for displacement constraint are shown in Figure 8.
- Combine the PSO algorithm directly to the response surface and evaluate the optimal design solution.

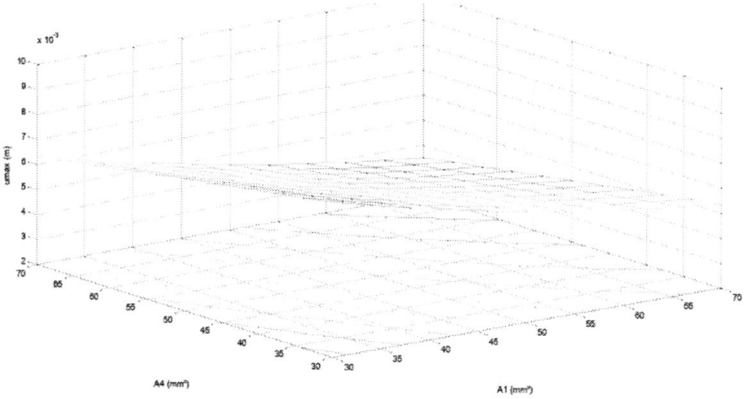

Figure 8. Response plot for maximum displacement

The results obtained for this problem using the Modified PSO Algorithm is given in Table 1.

Table 1. Design point from PSO algorithm

Parameter	Initial point	ANSYS First order	MPSO algorithm
A1 (mm)	60	26.85	27.05
A2 (mm)	55	25.94	25.34
A3 (mm)	55	24.86	24.14
A4 (mm)	50	24.54	24.91
Umax (m)	0.0025	0 .0085	0 .0089
Fobj (Kg)	3.402E7	1.466E7	1.451e+007

It is observed here that the results for the objective function of the MPSO optimization process compare well with those of ANSYS First order optimization process and the optimal solution is obtained in an optimized computation time.

6. Conclusion

In this paper, a parallelized version of the Particle Swarm Optimization algorithm is presented. A new concept of evolutionary neighborhoods is associated to the parallel model in order to improve the PSO performance. PSO is a stochastic metaheuristic based on population solutions. It searches for optimal

solutions based on the concepts of cooperation and neighborhoods. Many variations and improvements of classical PSO version have been suggested by adapting its parameters, but good algorithm acceleration is required with a parallelization approach.
In order to test the proposed PSO algorithm, we designed and applied the developed optimization procedure to electricity pylon example. The application problem studied is weight minimization of bars truss, by finding optimal cross-sectional areas of the truss members. The results obtained solving this optimization problem using the proposed PSO algorithm is then compared with the results from ANSYS first order conventional optimization technique.
The PSO algorithm was implemented sequentially, and then in parallel, based on the exchange between the subswarms. By testing the parallel version, and through the results, we find that our model shows remarkable efficiency in terms of reduced time and optimality convergence.
Finally, in the future we intend to study other variants of the suggested parallel model and also to test the program on other various real optimization problems.

7. References

[1] M. Pardalosa, E. Romeija, H. Tuyb: Optimization and Nonlinear Equations Recent developments and trends in global optimization. In: Journal of Computational and Applied Mathematics, Numerical Analysis 2000 Vol 124, pp. 209-228, 2000.

[2] J. Kennedy and R. Eberhart,: “Particle Swarm Optimization,” In: Proceedings of the IEEE International Joint Conference on Neural Networks, IEEE Press, vol. 8, no. 3, pp. 1943–1948, 1995.

[3] Y. Cooren, “Perfectionnement d'un algorithme adaptatif d'Optimisation par Essaim Particulaire. Applications en génie médical et en électronique”. Doctorat thesis, University of Paris 12 Val de Marne, France, 2008.

[4] K. E. Parsopoulos and M. N. Vrahatis: “Recent approaches to global optimization problems through particle swarm optimization”. In: Natural Computing: an international journal, 1(2-3), pp.235-306, 2002.

[5] M.E. Hyass and P. Hyass: “Good Parameters for Particle Swarm Optimization”. In : Laboratories Technical Report no. HL1001, 2010.

[6] B. Bochnek et P. Fory’s, “Structural optimization for post buckling behavior using particle swarms”. Struct Multidisc Optim. Pages 521-531, 2006.

[7] N. Elhami, R. Ellaia, M. Itmi: “Hybrid Evolutionary Optimization Algorithm MPSO-SA”, In: International Journal of Simulation and Multidisciplinary Optimimisation, Vol. 4, pp. 27- 32, 2010.

[8] P. Rabanal, I. Rodríguez and F. Rubio: “Parallelizing Particle Swarm Optimization in a Functional Programming Environment ». In Algorithms2014 : vol. 7, pp. 554–581, 2014.

[9] K. Byung-I and G. Alan, "Parallel asynchronous particle swarm optimization," International Journal For Numerical Methods In Engineering, vol. 67, pp. 578-595, 2006.

[10] J. Chang, S. Chu, J. Roddick and J. Pan : « A Parallel Particle Swarm Optimization Algorithm With Communication Strategies ». In :Journal of Information Science and Engineering, 2005.

[11] M.Zemzami, N.Elhami, M.Itmi and N.Hmina, “Parallèlisation de la Méthode PSO: Découpage de l’espace et traitement par lot des particules“. In: International Workshop on New Services and Networks (WNSN’16). Khouribga. Morocco. 2016.

[12] M.Zemzami, N.Elhami, M.Itmi and N.Hmina, “A New Parallel Approach For The Exploitation Of The Search Space Based On PSO Algorithm”. In: International Colloquium in Information Science and Technology (CIST’16). Tangier. Morocco. Indexed in Scopus 2016.

[13] M.Zemzami, N.Elhami, M.Itmi and N.Hmina, “Parallelization of the PSO algorithm on evolutionary neighborhoods”. In: International Conference on Modeling, Optimization and Simulation (MOSIM’16). Montréal. Canada. 2016.

[14] ANSYS Guide 2015, ANSYS Structural Analysis Guide, 2015.

Acknowledgments
This research is supported by « XTERM »: Complex Systems, Territories Intelligence and Mobility, co-financed by the European Union with the European regional development fund (ERDF) and Normandy Region.

Development and Application of Fiber Bragg Grating Clinometer

Guo Xin, Li Wen, Wang Wentao, Feng Xiaoyu

Changsha University of Science and Technology, Institute of Transportation Engineering, Hunan, China

563456009@qq.com

Abstract. Using FBG (fiber bragg grating) technology in clinometers can solve the technological problem facing by wireless transmission devices like big data transfer volume and poor stability, which has been receiving more and more attention. This paper discusses a new clinometer that is designed and transformed based on upgrading current clinometers, installing fiber grating strain gauges and fiber thermometers, and carrying out studies on such aspects as equipment upgrading, on-site setting, and data acquisition and analysis. In addition, it brings up the method of calculating displacement change based on wavelength change; this method is used in safety monitoring of the right side slope of Longyong Expressway ZK56+860 ~ ZK56+940 Section. Data shows that the device is operating well with a higher accuracy, and the slope is currently in a steady state. The equipment improvement and the method together provide reference data for safety analysis of the side slope.

1. Introduction

Slope clinometer is commonly used in engineering control. Traditional method is to use throw-in cable tester to measure the displacement number of clinometer tube[1], which is widely used. The advantage is mature technology and low cost while the disadvantage is that, the result is easy to be affected by climate and data acquisition is lagging. Monitoring technology has achieved a leapfrog development recently; GPS and optical fiber technology are used in slope monitoring[2]. GPS method has a wide application area. The advantage is easy data acquisition and can achieve remote operation; disadvantage is limited accuracy, instability, and high costs. While optical fiber technology, with the advantage of high accuracy, easy installation, and can achieve long-term long distance real time monitoring, has been using widely in geotechnical engineering. Comparing with other fiber optic sensors, FBG sensor has high sensitivity, and long service life. Optical grating with different pitch can be placed in the same optical fiber, to monitor temperature, stress, and strain at the same time[3,4]. It is feasible to combine FBG technology and traditional technology for measuring borehole deviation for which can both guarantee the accuracy of the data while achieving real time monitoring[5,6].

2 Designing and Installation of FBG Clinometer

2.1 Production of FBG Clinometer Sensor

The base material of the device is traditional PVC clinometer tube. Use slice-shaped FBG strainometer as monitoring element (element size is 1cm×6.5cm) and install it into the slide way, as graph 1[6], and group the element based on the wavelength shown on the element and the depth of each hole. After completing the installation inside each tube, put a temperature sensor to reduce the impact on wavelength detection caused by temperature variation; and using cable to connect each sensor.

2.2 Installation of the Clinometer

The drilling depth of clinometer hole is more than ten meters and has distinct diurnal temperature differences. Therefore, FBG temperature sensor should be installed in each hole to compensate the temperature[12,13]. To connect 2 to 3 temperature sensors in series into displacement sensor system based on the length of the slope to compensate the temperature.

The detailed procedure of clinometer connecting installation is:

① Put a voltage-type liquidometer into the clinometer tube in the bottom, and pierce cable into clinometer tube one by one, so as to prevent liquidometer from failing to be put in the right place because there are too many cables in the tube after finishing pasting the strainometer;

② Traditional clinometer tube is placed and installed one by one. But because there are too many cables inside the device, more attention should be paid to protect the cable while installing; one person for each tube, and install at the same time.

③ other steps are similar to the burial of traditional clinometer tube, which are: drilling (pore diameter of the drill is 10cm), down tube, discharging mud mixture by injecting water, sand filling in the hole, sealing and fixation. Binding all cables and numbering each tube after completing the burial.

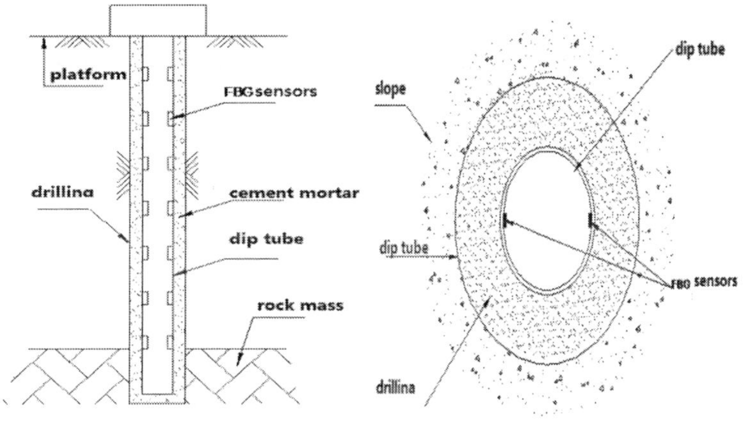

(a) longitudinal section (b) transverse section

Figure.1. Installation Drawing of Clinometer

3 FBG Signal Acquisition and Transmission

Signal acquisition is mainly divided into two processes: sensing and demodulation. Sensing is to modulate the parameter into the central wavelength of the sensor, and demodulation is to shift the reflected wave signal of the sensor into parameters. There are many methods to demodulate the signal, and this paper adopts modulable F-P filtering demodulation method. Considering the environment and condition of the slope are rather different during construction period and operating period, the signal acquisition and transmission plan is also divided into two modes; in construction period and in operating period.

3.1 The Acquisition and Transmission of ClinometerSignal during Construction Period

During the construction period, the slope has not been fully molded yet, and its reinforcement measure has not completed either. The platform has no conditions for laying cables and installing system and other perpetual transmission equipment. At this stage, the most commonly used way is manual data collection. The working process is as follows:

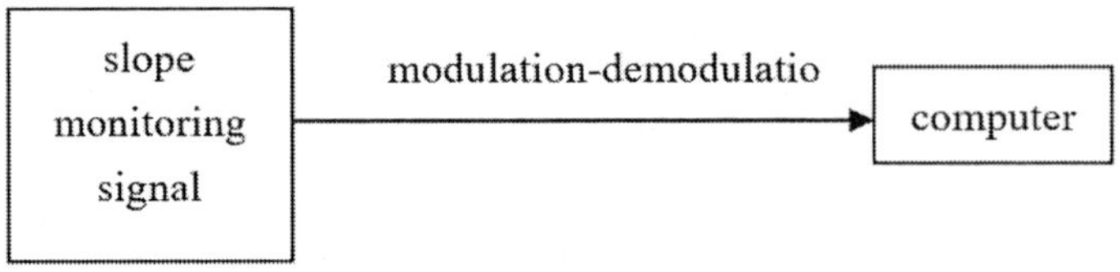

Figure.2. Construction Period Data Transmission Flow Chart

3.2 Signal Acquisition and Transmission during Operating period

During the operation period, the slope is comparatively stable and has conditions for laying long-term transmission equipment. Therefore, as for data collection, collection- demodulation-transmission method is adopted during operation period, namely, put a demodulator on the side of the slope, the wavelength signal of the sensor will then be demodulated into digital signal and transferred to the processing terminal in the slope maintenance institute.

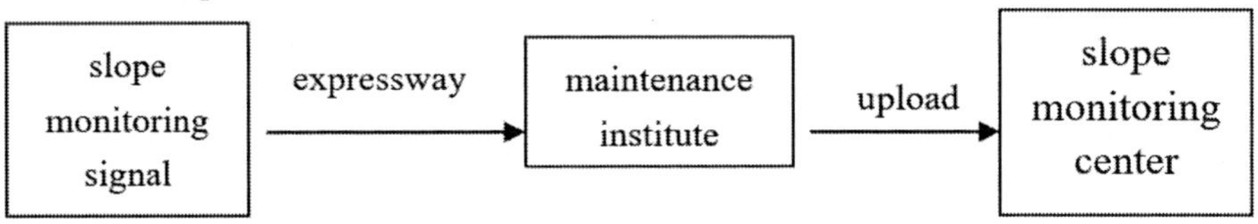

Figure.3. Operating Period Data Transmission Flow Chart

4 Engineering Application

4.1 Project Overview

The height of the highest place in the right side slope of Longyong Expressway ZK56+860 ~ ZK56+940 Section is 23 meters, which makes it a level 3 slope. It is mainly composed of clay and batt. The slope is located near the Danongche fault zone in Longshan County with broken rocks and poor stability. This FBG clinometer is buried in July 2014, and started monitoring since then. The direction inward the slope is defined as positive, and the outward the slope as negative.

4.2 Data Processing

FBG wavelength data cannot directly reflect the displacement situation of the slope; it can only reflect the change of grid pitch. It is necessary to do the conversion; to get the deflection of different point locations of the clinometer tube through strain calculation, and do the conversion to get slope displacement.

Using the difference method to calculate the deflection of the clinometer; the stress equation and stress-strain relation equation of FBG root position clinometer can be expressed as:

$$\frac{1}{h^2}(f_{x+2h} - 2f_{x+h} + f_x) = \frac{1}{R}\frac{MR}{EI} = \frac{\varepsilon_i}{R} \quad (1)$$

The matrix expression is:

$$\frac{R}{h^2}\begin{bmatrix} 1 & -2 & 1 & 0 & \cdots & \cdots & 0 \\ 0 & \ddots & \ddots & \ddots & \ddots & \ddots & \vdots \\ \vdots & \ddots & \ddots & \ddots & \ddots & \ddots & \vdots \\ \vdots & \ddots & \ddots & \ddots & \ddots & \ddots & \vdots \\ \vdots & \ddots & \ddots & \ddots & \ddots & \ddots & 0 \\ 0 & \cdots & \cdots & 0 & 1 & -2 & 1 \end{bmatrix} \begin{Bmatrix} f_x \\ f_{x+h} \\ \vdots \\ f_{x+(n-1)h} \end{Bmatrix} = \begin{Bmatrix} \varepsilon_1 \\ \varepsilon_2 \\ \vdots \\ \varepsilon_n \end{Bmatrix} \quad (2)$$

In the matrix: and are deflections of the fixed end. , n refers to the number of sensor, and the maximum in this research is 7) is the deflection of the place of the ith optical brag grating sensor. h refers to the distance between two neighboring optical brag grating sensors, and in this research, the distance is 2m. R refers to the inner diameter of root position cable tester, and in this research, the specification of PVC clinometer tube is 0.58m. refers to the strain value which is monitored by the ith optical brag grating sensor counting from the bottom
Assuming, and the bottom of the clinometer is buried under the rock stratum for 0.5m, which can be regarded as the fixed end. And are the deflection of the virtual point on the fixed end. Therefore, and we can get from the equation that:

$$\frac{R}{h^2}\begin{bmatrix} 1 & -2 & 1 & 0 & \cdots & \cdots & 0 \\ 0 & \ddots & \ddots & \ddots & \ddots & \ddots & \vdots \\ \vdots & \ddots & \ddots & \ddots & \ddots & \ddots & \vdots \\ \vdots & \ddots & \ddots & \ddots & \ddots & \ddots & \vdots \\ \vdots & \ddots & \ddots & \ddots & \ddots & \ddots & 0 \\ 0 & \cdots & \cdots & 0 & 1 & -2 & 1 \end{bmatrix} \begin{Bmatrix} 0 \\ 0 \\ f_{2h} \\ \vdots \\ f_{(n+1)h} \end{Bmatrix} = \begin{Bmatrix} \varepsilon_1 \\ \varepsilon_2 \\ \vdots \\ \varepsilon_n \end{Bmatrix} \quad (3)$$

In formula 3, the first and second column of the matrix can be neglected, and the equation can be simplified as:

$$\frac{R}{h^2}\begin{bmatrix} 1 & 0 & 0 & 0 & \cdots & \cdots & 0 \\ -2 & 1 & 0 & \ddots & \ddots & \ddots & \vdots \\ 1 & -2 & \ddots & \ddots & \ddots & \ddots & \vdots \\ 0 & 1 & \ddots & \ddots & \ddots & \ddots & \vdots \\ \vdots & \ddots & \ddots & \ddots & \ddots & \ddots & 0 \\ 0 & \cdots & \cdots & 0 & 1 & -2 & 1 \end{bmatrix} \begin{Bmatrix} f_{2h} \\ f_{3h} \\ \vdots \\ f_{(n+1)h} \end{Bmatrix} = \begin{Bmatrix} \varepsilon_1 \\ \varepsilon_2 \\ \vdots \\ \varepsilon_n \end{Bmatrix} \quad (4)$$

The above matrix is reversible. By finding the inverse matrix, one can figure out the relation between displacement and the strain of the measuring point. As follows:

$$\begin{Bmatrix} f_{2h} \\ f_{3h} \\ \vdots \\ f_{(n+1)h} \end{Bmatrix} = \frac{h^2}{R}\begin{bmatrix} 1 & 0 & 0 & 0 & \cdots & \cdots & 0 \\ -2 & 1 & 0 & \ddots & \ddots & \ddots & \vdots \\ 1 & -2 & \ddots & \ddots & \ddots & \ddots & \vdots \\ 0 & 1 & \ddots & \ddots & \ddots & \ddots & \vdots \\ \vdots & \ddots & \ddots & \ddots & \ddots & \ddots & 0 \\ 0 & \cdots & \cdots & 0 & 1 & -2 & 1 \end{bmatrix}^{-1} \begin{Bmatrix} \varepsilon_1 \\ \varepsilon_2 \\ \vdots \\ \varepsilon_n \end{Bmatrix} \quad (5)$$

That is:

$$\begin{Bmatrix} f_{2h} \\ f_{3h} \\ \vdots \\ f_{(n+1)h} \end{Bmatrix} = \frac{h^2}{R}\begin{bmatrix} 1 & 0 & 0 & 0 & \cdots & \cdots & 0 \\ 2 & 1 & 0 & \ddots & \ddots & \ddots & \vdots \\ 3 & 2 & \ddots & \ddots & \ddots & \ddots & \vdots \\ 4 & 3 & \ddots & \ddots & \ddots & \ddots & \vdots \\ \vdots & \ddots & \ddots & \ddots & 2 & 1 & 0 \\ n & n-1 & \cdots & 4 & 3 & 2 & 1 \end{bmatrix} \begin{Bmatrix} \varepsilon_1 \\ \varepsilon_2 \\ \vdots \\ \varepsilon_n \end{Bmatrix} \quad (6)$$

This shows that, by model simplification and assuming the clinometer, one can establish the relation between strain and displacement. In formula 6, except the strain value, which is acquired by calculating the data collected from FBG demodulator, the rest are constant. Therefore, the treating method of the data can also be used as the computing method of the data collected by the device.

4.3 Displacement Analysis

By analyzing above equation one can get relevant data of this slope, and after processing, one can get its displacement-time graph. As shown below:

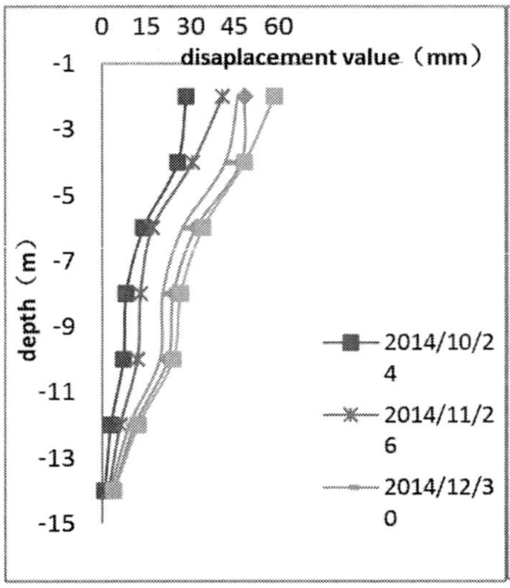

(a) 1# hole displacement-time changing curve

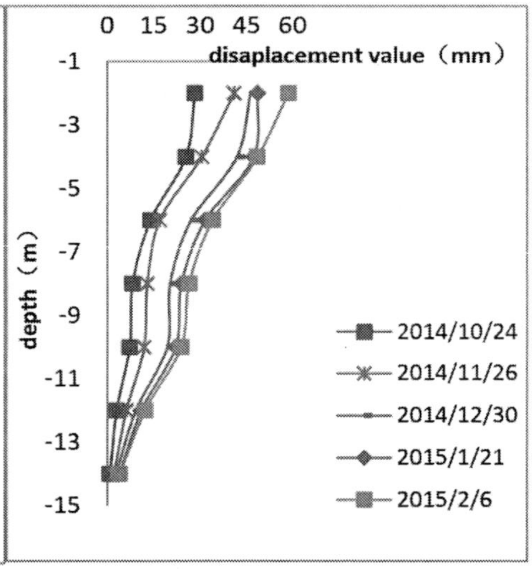

(b) 2# hole displacement-time changing curve

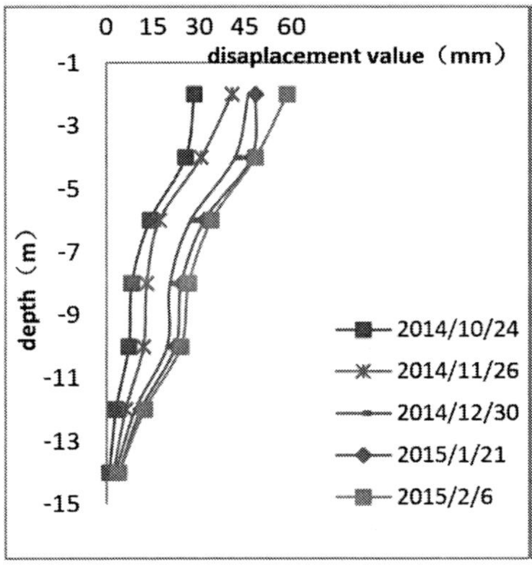

(c) 3# hole displacement-time changing curve

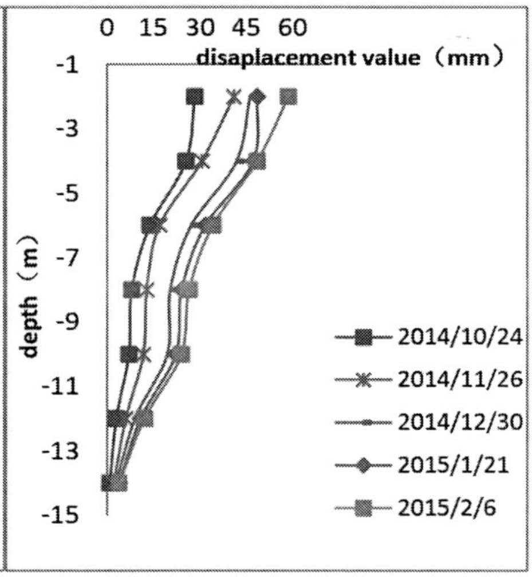

(d) 4# hole displacement-time changing curve

Figure.4. Displacement-time Changing Curve

The displacement-depth curves of each measured holes are V-shaped, and the upper displacement is larger and the lower displacement is smaller. The curves of 1# and 4# holes have obvious convex points at the position 3~4m below the platform, which shows that the upper displacement, which is more than 3m, of the above two holes is larger than that of the lower displacement. The maximum monthly average displacement of the single hole of the slope is 15~17mm, which is located in 2# and 3# holes which are at the end of two sides of the slope, and the #1 and 4# holes in the middle of the slope have smaller accumulative displacement change. According to the data and data processing method of the new device, the displacement curve is basically the same as the slope displacement trend, the accuracy of the measured data is higher than that of the traditional device, and the performance is improved obviously.

5 Conclusion

(1)This paper introduced the design, transform, inbuilt, signal transmission and data processing method of a clinometer, which is based on FBG technology. It offers new plan for monitoring the stability of the slope, and has obvious popularization using value.

(2)By monitoring the right side of Longyong Expressway ZK56+860 ~ ZK56+940 Section, the clinometer provides data reference for maintenance management in operating period. Comparing with traditional clinometer, the new one is more accurate, convenient in data collection; and it can get more data.

Reference

[1] Ma Quanzhen, Zhang Baohua, Application of Borehole Clinometer in Slope Monitoring[J]. Journal of Changzhou Institute of Technology, 2005(S1).

[2] Shen Qiang, Study on Monitoring and Prediction of Layered Rock Slope Stability of Highway in Mountainous Region[D]. Graduate University of the Chinese Academy of Sciences, Wuhan Institute of Rock and Soil Mechanics, 2007.

[3] Sui Haibo, Shi Bin, Zhang Dan, Wang Baojun, Wei Guangqing, Piao Chunde, REVIEW ON FIBER OPTIC SENSOR-BASED MONITORING TECHNIQUES FOR GEOLOGICAL AND GEOTECHNICAL ENGINEERING[J]. Journal of Engineering Geology, 2008(01).

[4] Song Zhen, Wang Baojun, Shi Bin, Zhang Dan, Sui Haibo, Li Haitao, Comparative Analysis of Fiber Monitoring and Numerical Simulation on the K3 Slope of Guangdong-Jiangxi Expressway[J]. Journal of Disaster Prevention and Mitigation Engineering, 2009(04).

[5] Sui Haibo, Shi Bin, Zhang Dan, Wang Baojun, Wei Guangqing, Piao Chunde, STUDY ON DISTRIBUTED OPTICAL FIBER SENSOR-BASED MONITORING FOR SLOPE ENGINEERING[J]. Chinese Journal of Rock Mechanics and Engineering, 2008(S2).

[6] Pei Huafu, Yin Jianhua, Zhu Honghu, Hong Chengyu, Fan Youhua, IN-SITU MONITORING OF DISPLACEMENTS AND STABILITY EVALUATION OF SLOPE BASED ON FIBER BRAGG GRATING SENSING TECHNOLOGY[J]. Chinese Journal of Rock Mechanics and Engineering, 2010(08).

[7] Peng Bei, Meng Lijun, Research and Development on Distributed Fiber Bragg Grating Sensors[J]. Software Guide, 2012(06).

[8] Yang Xing, Hu Jianming, Dai Teli, Principle and Typical Current Applications of Fiber Grating Sensors[J]. Journal of Chongqing Normal University(Natural Science), 2009(04).

[9] Pang Dandan, Investigation on the Novel Fiber Bragg Grating Sensing Technology[D]. Shandong University, 2014.

[10] Zhu Honghu, Yin Jianhua, Jin Wei, Gu Dongming, Health monitoring of foundations using fiber Bragg grating sensing technology[J]. China Civil Engineering Journal, 2010(06).

[11] Chen Pengchao, Li Jun, Liu Jianping, Jin Shijiu, Monitoring technology of pipelines using fiber bragg grating and its application in landslide areas[J]. Chinese Journal of Geotechnical Engineering, 2010(06).

[12] Hu Zhixin, Wang Zhenwu, Ma Yunbin, Zhang Jun, Soil pressure sensor based on temperature compensation FBG[J]. Journal of Applied Optics, 2010(01).

[13] Zhao Tao, Monitoring Principle and Method of Transformer Internal Temperature Based on the FBG Temperature Sensing[D]. Chongqing University, 2008.

[14] Yao Hongxu, Wei Bingxu, Qualitative analysis and estimation of inclination survey data curve in slope slip monitoring[J]. Engineering Construction, 2011(03).